Advances in Industrial Control

Other titles published in this series:

John Pittner · Marwan A. Simaan

Tandem Cold Metal Rolling Mill Control

Using Practical Advanced Methods

 Springer

Dr. John Pittner
University of Pittsburgh
Dept. Electrical & Computer
Engineering
Benedum Hall 348
15260 Pittsburgh Pennsylvania
USA
jrpst16@pitt.edu

Prof. Marwan A. Simaan
University of Central Florida
Department of Electrical Engineering &
Computer Science
Central Florida Blvd. 4000
32816-2362 Orlando Florida
USA
simaan@eecs.ucf.edu

ISSN 1430-9491
ISBN 978-0-85729-066-3   e-ISBN 978-0-85729-067-0
DOI 10.1007/978-0-85729-067-0
Springer London Dordrecht Heidelberg New York

Springer is part of Springer Science+Business Media (www.springer.com)

Advances in Industrial Control

Professor (Emeritus) O.P. Malik
Department of Electrical and Computer Engineering
University of Calgary
2500, University Drive, NW
Calgary, Alberta
T2N 1N4
Canada

Professor K.-F. Man
Electronic Engineering Department
City University of Hong Kong
Tat Chee Avenue
Kowloon
Hong Kong

Professor G. Olsson
Department of Industrial Electrical Engineering and Automation
Lund Institute of Technology
Box 118
S-221 00 Lund
Sweden

Professor A. Ray
Department of Mechanical Engineering
Pennsylvania State University
0329 Reber Building
University Park
PA 16802
USA

Professor D.E. Seborg
Chemical Engineering
3335 Engineering II
University of California Santa Barbara
Santa Barbara
CA 93106
USA

Doctor K.K. Tan
Department of Electrical and Computer Engineering
National University of Singapore
4 Engineering Drive 3
Singapore 117576

Professor I. Yamamoto
Department of Mechanical Systems and Environmental Engineering
The University of Kitakyushu
Faculty of Environmental Engineering
1-1, Hibikino,Wakamatsu-ku, Kitakyushu, Fukuoka, 808-0135
Japan

Series Editors' Foreword

The series *Advances in Industrial Control* aims to report and encourage technology transfer in control engineering. The rapid development of control technology has an impact on all areas of the control discipline. New theory, new controllers, actuators, sensors, new industrial processes, computer methods, new applications, new philosophies, new challenges. Much of this development work resides in industrial reports, feasibility study papers and the reports of advanced collaborative projects. The series offers an opportunity for researchers to present an extended exposition of such new work in all aspects of industrial control for wider and rapid dissemination.

In the 1970s, both the Editors of the *Advances in Industrial Control* monograph series belonged to the Industrial Automation Group at Imperial College of Science and Technology (as it was then known) in London. Under the leadership of Professor Greyham F. Bryant, it was this Group that did so much to create a new paradigm for the study and control of tandem cold rolling mills. It was there that the successful and influential non-interacting control networks concept was formalised and promoted for industrial implementation. The monograph *The Automation of Tandem Mills* (1973) by Professor Bryant and many of the Group's staff is often cited as a seminal reference for all subsequent research and development in this field.

Since the publication of the 1973 monograph describing the approach of the Industrial Automation Group, the Series Editors have not seen many books published on tandem cold rolling mill control. Consequently, this *Advances in Industrial Control* monograph, *Tandem Cold Metal Rolling Mill Control: Using Practical Advanced Methods* by John Pittner and Marwan A. Simaan is making a very timely and welcome appearance. Recent years have seen system methods undergo significant advances in the design of control systems for multivariable systems with significant interactions. There have also been rapid advances in robust control systems design methods involving a deeper understanding of how to treat uncertainty in models and disturbances. In addition there has been a determined effort within the control community to develop and to use nonlinear control techniques in industrial applications. The tandem cold rolling mill, being a multivariable and highly nonlinear process is a suitably challenging process on which to assess some of the new control approaches, and John Pittner and Marwan Simaan tackle this very challenge in their new monograph.

The structure of the monograph follows the usual systems approach firstly concentrating on describing, analysing, and modelling the tandem mill process (Chapter 2). Following this is a useful discussion of "conventional" non-interacting control along with the later variants of these methods that have their roots in the 1970s paradigm (Chapter 3). Advanced control is developed in two stages. In Chapter 4, two techniques, H_∞ loop shaping, and observer-based methods are studied and simulation results given. In Chapter 5, the authors advance their case for the use of the state-dependent matrix Riccati control approach. The investigations in Chapters 3–5 are accompanied by insightful discussion and conclusion sections. To augment and enhance the initial process descriptions of Chapter 2, the authors present more detail of the motor drive systems of the process in the final chapter of the monograph.

As is well known, in many cases improved control systems are often the only economical way to enhance the performance of industrial installations involving large capital investment. Similarly, a new control system may be the only way to revitalise and extend the operational life of ageing industrial plant. This monograph will be essential reading for those professionals and technicians involved in the operation of tandem mills, or in the development of tandem mill control systems design and technology.

A wider readership in the steel industry, in the control systems technology field, and in the academic community will find the monograph gives an introduction to the process and control of tandem cold rolling mills, and provides a useful case-book study to compare "conventional" control methods versus the potential of advanced modern control, for a significant and technologically challenging industrial process.

Industrial Control Centre   M.J. Grimble
Glasgow   M.A. Johnson
Scotland, UK
2010

Preface

The purpose of this monograph is twofold. Firstly, it is intended to introduce the augmented state-dependent Riccati equation method as a novel means of advanced control for the improvement of the quality of the final product of the tandem cold metal rolling process. Second, as background for the presentation of the state-dependent Riccati equation method, it provides an overall review of some of the basic areas in the control of the tandem cold metal rolling process and some of the more conventional and advanced methods for control in certain of these areas. It is considered that the book will serve as a means of presenting to control practitioners and process engineers in the associated areas of metals process control a novel, viable, and practical technique for the control of tandem cold metal rolling. On the other hand, for control theoreticians the material presented will provide insight into some of the more practical issues associated with the control of the process, many of which are quite challenging, important to improvement in the quality of the final product, and by no means trivial. The material presented in this monograph is the result of about 6 years of work at the University of Pittsburgh related to the application of the augmented state-dependent Riccati equation method to tandem cold rolling. As can be seen in what is presented in the following chapters the results of the work to date have been quite successful.

A major issue in achieving an improvement in the quality of the final output is the improvement in the control of centerline thickness and interstand tension. While there are many conventional methods presently in use for control of the process in these areas, most of them are based on single-input-single-output (SISO) control structures which limits their capability for improvement, even though such conventional methods have done a reasonably good job in producing a product that is of good quality, and being generally user-friendly and fairly easy to tune which has kept commissioning times within reasonable bounds. However, because of their SISO-type structures most of the conventional techniques presently in use are limited in their capability to improve performance because of their limitations in handling the dynamic interactions between the many variables in the mill, which is a large nonlinear system with speed-dependent time delays, a broad range of both external and internal uncertainties, and significant disturbances, many of which can change during operation. Moreover, the present trend is for tandem cold rolling mills to be coupled to continuous pickling processes so that the entire operation is continuous. This puts an added requirement on the controller of the tandem cold rolling process to successfully handle very rapid changes in the thickness, hardness, and width of the products being processed, as the transition at a mill stand between two coupled products occurs in milliseconds at nominal operating speeds. In addition, the controller must offer a structure that is user-friendly to design and commissioning personnel most of whom have a limited background in advanced control theory. Such a controller structure is crucial to the reduction in control engineering design costs and the reduction of commissioning efforts to assure a timely and cost effective transition from commissioning to production.

In response to these requirements we have investigated the applicability of several possible advanced control methods as viable candidates for control of the mill. While there are certain advantages and disadvantages for each of the candidates, the one that appeared to be most promising was the state-dependent Riccati equation technique which has many features that are desirable for control of the tandem cold rolling process. During the investigation of this method, it was discovered that very simple augmentations to the structure of the basic controller enhanced the desirable features of this technique to provide an overall control system that improves performance while considering typical uncertainties and disturbances, and is user-friendly to design and commissioning personnel.

The content of this book is divided into six chapters. Chapter 1 provides an introduction to and a brief history of the tandem cold metal rolling process. In Chapter 2 a process model suitable for control development is presented and data is included that confirms its verification. Chapter 3 is a brief review of several conventional methods of control of centerline thickness and tension, and for completeness briefly touches on other areas such as control of flatness and threading. Chapter 4 presents other methods of advanced control and discusses some of their advantages and disadvantages. In Chapter 5 the augmented state-dependent Riccati technique is introduced and the results of simulations are given which demonstrate significant improvements in performance when compared to well-performing conventional methods. Chapter 6 supplements the previous chapters on the control of strip thickness and tension by providing some introductory material on the control of motors and drives to give a more complete picture of the various areas involved in the overall control of the mill.

It is considered that the results of the work presented in this book will be of benefit to associated control practitioners, process engineers, and theoreticians so that the advanced technique described herein ultimately can be applied to actual operating mills to realize a significant improvement in the control of an important industrial process.

Pittsburgh, PA   John Pittner
Orlando, FL   Marwan A. Simaan
December 2010

Acknowledgements

For this work the authors wish to express their gratitude for support in part by the National Science Foundation under grant 0951843 and in part by the Pennsylvania Infrastructure Technology Alliance, a partnership of Carnegie Mellon, Lehigh University, and the Commonwealth of Pennsylvania's Department of Community and Economic Development (DCED).

Contents

Notation

Unless noted otherwise in the text the notation is as follows. The notation for individual variables representing physical quantities is as noted specifically in the associated text.

R	The field of real numbers
$\in$	Belongs to
$\equiv$	Defined as
A'	Transpose of matrix A
A^{-1}	Inverse of matrix A
I	Identity matrix
$\|x\|$	Euclidean norm of vector x
$\|A\|_{\infty}$	Infinity norm of matrix A
$\in C^k$	The elements of a vector or matrix have continuous partial derivatives through order k
∇	Gradient. The computation of matrix gradients is as noted in Sects. 5.6.2 and 5.6.4.

Chapter 1
Introduction

1.1 Objectives

The tandem rolling of cold metal strip such as steel is a complex nonlinear multivariable process the control of which presents a significant engineering challenge. The current technology for control of this process generally relies on a structure developed in the UK early in the 1970s by G. F. Bryant [1]. In this technique the undesirable effects of interactions between the many process variables are partially mitigated by single-input-single-output (SISO) and single-input-multi-output (SIMO) control loops operating on selected variables. Additionally, the overall control problem is decomposed into several separate problems with the objective of providing independent adjustment of strip tension and thickness anywhere in the mill. This structure and variations of it [*e.g.* 2, 3] have been effective in producing an acceptable product. However, it is recognized both by control theoreticians and practitioners that applications of other design techniques could result in improvements in performance, particularly in robustness to disturbances and uncertainties. Various advanced methods [*e.g.* 4, 5] based on well-established techniques have been proposed and simulated in academia, and a few have been implemented in actual practice. Many of these advanced methods have resulted in some improvements, but also have some significant shortcomings resulting mostly from the complexities of the control methods. Among these are difficulties in tuning during commissioning by personnel who usually are unfamiliar with advanced control techniques, complexities in the controller design methods, and the necessity for development of a linearized process model. Consequently, there is a need for a better approach.

As a step toward fulfilling this need, an objective of this book is to investigate several approaches to the control of the tandem cold metal rolling process. This investigation includes both conventional and advanced methods that have been proposed and simulated as well as those that have been actually implemented, with the goal of identifying one or more particular methods that offers a significant advantage. The work covers both theoretical and applied aspects, provides comments on the various approaches, and to the extent practical compares the various approaches to the results of actual installations. The success of a particular method

J. Pittner and M.A. Simaan, *Tandem Cold Metal Rolling Mill Control*,
Advances in Industrial Control, DOI 10.1007/978-0-85729-067-0_1,
© Springer-Verlag London Limited 2011

is judged on its capability to improve performance, to provide a controller that is easy to implement at commissioning, to enhance the use of physical intuition in a design process that is structured to be user-friendly, and does not require a linearized model. Of course, not every conceivable method can be evaluated, nor can a review be exhaustive in every detail. Thus what is considered are the salient features of those advanced methods based on well-established techniques which have received attention in the literature and appear to offer the potential for significant improvement over conventional methods. Also included are the features of well-performing conventional methods that are typical for control of tandem cold rolling mills and for which performance data are readily available.

To provide some background to aid in understanding of what is presented in the chapters that follow, this chapter presents a brief historical overview of tandem cold metal rolling and a description of the basic characteristics of the more common tandem cold rolling processes, *i.e.* stand-alone tandem cold rolling and continuous tandem cold rolling.

1.2   Historical Background

Some of the earliest known rolling of metals was somewhere around the fourteenth century where very small rolls were used to flatten cold metals such as gold, silver, or lead that were used in jewelry or works of art. In 1480 Leonardo da Vinci describes in his notes machines for the cold rolling of sheets and bars, but it is unlikely that these mills were ever constructed. Toward the last half of the sixteenth century the cold rolling of metal, particularly lead, began to take on more importance as lead began to be used for roofing and other applications such as organ pipes.

The seventeenth century saw the hot rolling of bars of ferrous materials into thin sheets throughout Europe and especially in Germany, with applications later in that century in England and Wales. These mills were single stand mills. It was not until the eighteenth century that metal was rolled in successive stands arranged as a tandem mill, the first recorded of which was for the hot rolling of wire rods in England in 1766, and in 1798 a patent was issued for a tandem mill for rolling iron plates and sheets.

The size of the mills and of their rolled products increased significantly during the nineteenth century with the advent of the industrial revolution, which brought about a tremendous increase in the demands for iron and steel. Most mills prior to the middle of the nineteenth century were two-high (Figure 1.1). However, at about the middle of the century a three-high mill was introduced in Great Britain by R. B. Roden and later improved upon by Lauth who used a middle roll with a lesser diameter than the two larger backup rolls. This arrangement offered greater productivity with a significant reduction in the power required. The four-high arrangement was introduced in 1872 in England and slightly earlier in Germany for rolling rails and beams.

Fig. 1.1 Cold mill
arrangements

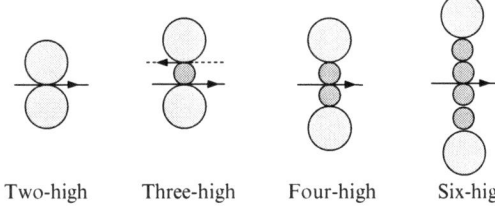

Two-high   Three-high   Four-high   Six-high

The Lauth three-high arrangement with the smaller diameter middle roll provided the driving force for the cold rolling of steel as a successful production process. The successful commercialization of this arrangement occurred mostly in the United States in Pittsburgh, Pennsylvania by the American Iron and Steel Company, which was later acquired by the Jones and Laughlin Steel Corporation. The first four-high mill for the cold rolling of steel was used experimentally by Allegheny Ludlum Steel Corporation in 1923.

The first known tandem cold rolling of steel strip was in 1904 at the West Leechburg Steel Company also near Pittsburgh, which was a two-high four-stand mill. However, real tandem cold mill operation with strip tension between the stands occurred around 1915 at the Superior Steel Company in Pittsburgh, and in 1926 the first two-high four-stand tandem cold mill began operation in the Butler Pennsylvania plant of the American Rolling Company. The 1930s saw significant advancements in the maximum widths of cold rolled strip from about 890 mm in 1935 to as much as 2,290 mm in 1940. Around the time of World War II and shortly thereafter most of the mills in use were four-stand, with five or six-stand mills coming into predominant usage in the early 1960s.

Considerable advancements in rolling technology were made in the 1970s, and these resulted in improved surface finishes, tighter dimensional tolerances, and higher rolling speeds. Factors that contributed to this progress were larger and faster mill designs, improvements in the mill rolls and housing, advances in variable speed drive technology, enhancements in the instrumentation systems, and the maturing of computer control. More recent developments are fully continuous tandem cold mills, where strip accumulators are utilized for storage during coil changes, and fully continuous mills directly coupled to continuous pickling process lines.

Some other modern advances include the replacement of motorized screw-downs with hydraulic cylinders, fully automatic thickness control systems, improvement in speed and tension control systems with mill exit speeds on the order of 1,500–2,100 m/min being attained, and improvements in the control of the flatness and width of the strip, all of which contributed to improvements in productivity and in the yield and quality of the mill output. Additionally, six-high arrangements (Figure 1.1) have recently come into use for improvement in correcting a wider range of flatness defects in the strip. More detail regarding the historical background of metal rolling can be found in Roberts [6].

1.3 Process Overview

The tandem cold rolling of steel strip is one process in a sequence of processes performed to convert raw materials into a finished product. The cold rolling process occurs after the hot rolling process wherein steel slabs are heated in a furnace, or produced in a continuous hot metal casting operation, and then rolled into coils of reduced thickness suitable for further processing. After hot rolling and prior to cold rolling, the hot rolled material undergoes a pickling process wherein the coiled strip is unwound and passed through an acid bath to remove the oxides formed during hot rolling. Just prior to recoiling, oil is applied to the strip to prevent rusting, eliminate damage due to scraping of adjacent coil wraps, and in certain instances to act as a partial lubricant for the first stand of the tandem cold mill. The cold rolling process then provides an additional reduction to produce thinner material since the reduction in thickness in the hot rolling process generally is limited to about 1.25 mm. In addition, cold rolling is done for one, or both, of the following: (1) to improve the surface finish, and (2) to produce mechanical properties in the strip which make it suitable for the manufacture of various products such as the automated making of cans.

In a typical stand-alone 5-stand tandem cold mill (Figure 1.2), the strip is passed through five pairs of independently driven work rolls, with each work roll supported by a backup roll of larger diameter. Figure 1.3 depicts a typical four-high mill stand arrangement.

As the strip passes through the individual pairs of work rolls, the thickness is successively reduced. The reduction in thickness is caused by very high compression stress in a small region (denoted as the roll gap, which for purposes of this work is also referred to as the roll bite) between the work rolls. In this region the metal is plastically deformed, and there is slipping between the strip and the work roll surface. The necessary compression force is applied by hydraulic rams, or in

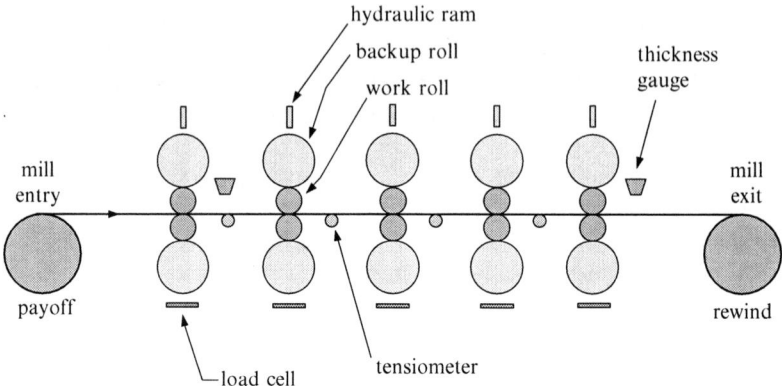

Fig. 1.2 Five-stand tandem cold mill, stand-alone configuration

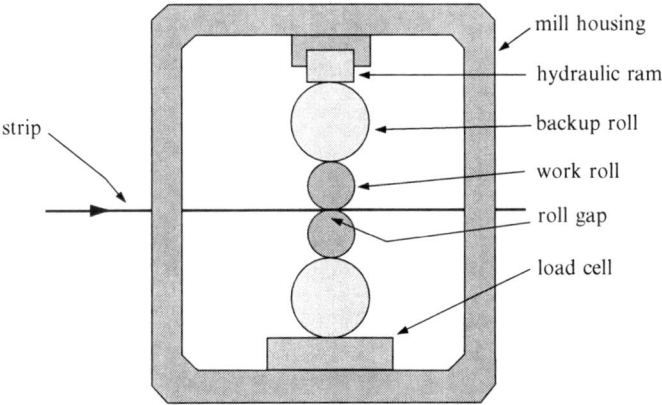

Fig. 1.3 Typical mill stand arrangement

many older mills by a screw arrangement driven by an electric motor. The energy required to achieve the reduction in strip thickness causes a temperature rise at the roll gap which is reduced considerably by the cooling effects of air and rolling solution (lubricant) as the strip travels between the stands.

Mill instrumentation generally consists of sensors to measure roll force at each stand, interstand strip tension force, strip thickness at the exit of the first and last stands, work roll speeds, roll gap actuator (hydraulic ram) positions, and in some instances the actual strip speed. Prior to rolling, work roll speed and roll gap position actuator references are calculated based on expected steady-state mill behavior.

During the threading process in a stand-alone mill the strip is successively introduced into the mill stands at low speed. After the last stand is threaded, the mill is accelerated to the desired operating speed (run speed). Near the end of the coil, the mill is decelerated to a reduced speed for de-threading and set-up for the next coil.

In the case of a continuous mill (Figure 1.4), the strip is fed from an upstream process, usually a continuous pickling process, through storage devices so that rolling is not interrupted for a coil change as in the stand-alone case. At the entry of this process, the strip of an incoming coil is welded to the strip of the coil being processed. As the weld exits the upstream process and approaches the tandem cold mill, generally the mill speed is lowered to reduce the likelihood of strip breakage during the weld passage, and to be within the cutting range of the shear. When the weld exits the mill and is at the shear a cut is made for a flying transfer of the strip to the available rewind, with the pinch roll closing to maintain tension during the transfer. When tension is established at the rewind, the pinch roll is opened and the mill speed is increased to the desired operating speed. Generally process monitoring for the continuous arrangement is essentially the same as in the stand-alone case, except with an additional strip thickness measurement at the mill entry.

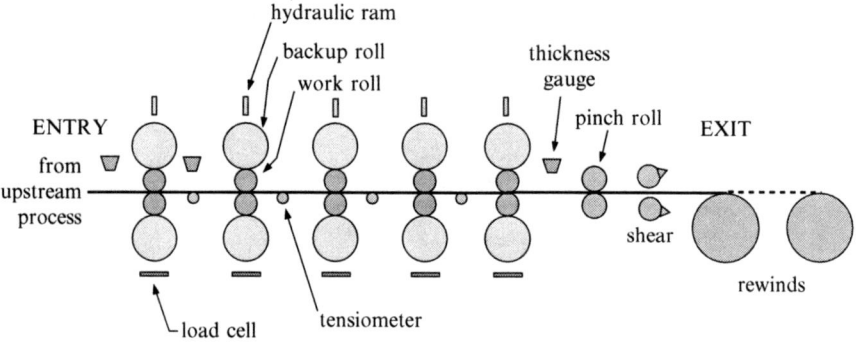

Fig. 1.4 Five-stand tandem cold mill, continuous configuration

After cold rolling in either a continuous or stand-alone operation, the strip is cleaned and annealed to restore its formability, which was reduced by an increase in hardness and a decrease in ductility caused by the cold reduction process.

1.3.1 Mill and Strip Data

In both the stand-alone and continuous configurations, the functions of different types of tandem cold rolling mills sometimes are identified based on the product being produced. Examples are sheet mills and tin mills, with "tin" implying that the rolled product can be processed further to be thinly coated with tin for ultimate use in a tinplate application such as the manufacturing of food cans and shipping containers. However, often a particular mill configuration is designed for sheet, tin, and other applications, as so-called tin mills are not solely committed to rolling only product that will be subsequently plated with tin. The distinction between sheet and tin essentially is based on the thickness and width of the product being processed. Table 1.1 presents some ranges of values for sheet and tin applications, as part of the mill and strip data.

Table 1.2 presents the thickness reduction patterns and interstand tensions of three production schedules [1] for a four-high, five-stand tandem mill rolling mild steel. The operator has the capability of slightly adjusting individually both the exit thicknesses and the interstand tensions during actual operation. The slight adjustment in thickness allows the shifting of the individual stand rolling loads during rolling. It should be noted that the data presented in Tables 1.1 and 1.2 are intended only to be illustrative of actual installations, and should be used only to give some rough feel for actual practice. The data for a particular installation will depend on the specific mill design, the type of product being processed, and the specific operational practices peculiar to the installation.

Table 1.1 Mill and strip data

Parameter	Dimension
Distance between stands	3.6–4.9 m
Work roll diameter	405–660 mm
Backup roll diameter	1,065–1,575
Roll face (sheet)	1,370–2,335
Roll face (tin)	890–1,370
Rolling force	20,000–40,000 kN
Overall reduction (sheet)	40–85%
Overall reduction (tin)	85–90%
Strip width (sheet)	460–2,035 mm
Strip width (tin)	610–915
Input strip thickness (sheet)	2.0–6.4
Output strip thickness (sheet)	3.6–6.1
Input strip thickness (tin)	1.8–2.5
Output strip thickness (tin)	0.18–0.45

Table 1.2 Production schedules for tandem cold rolling mill

Parameter	Production schedule number		
	1	2	3
Mill entry thickness, mm	3.56	2.36	1.78
Exit thickness, stand 1	2.95	2.01	1.22
Exit thickness, stand 2	2.44	1.52	0.79
Exit thickness, stand 3	2.01	1.22	0.56
Exit thickness, stand 4	1.68	0.97	0.38
Exit thickness, stand 5	1.58	0.91	0.36
Tension stress, mill entry, kN/mm^2	0.0	0.0	0.0
Tension stress, stands 1,2	0.086	0.091	0.097
Tension stress, stands 2,3	0.088	0.094	0.105
Tension stress, stands 3,4	0.089	0.097	0.119
Tension stress, stands 4,5	0.092	0.102	0.138
Tension stress, mill exit	0.028	0.028	0.028

1.3.2 Advantages of Continuous Tandem Cold Rolling

The arrangement for continuous tandem cold rolling wherein a tandem cold rolling mill is coupled to an upstream pickling process has several advantages over the stand-alone process. The most significant of these is the increase in productivity realized because of not having to interrupt the rolling process to change coils. As an example, for a stand-alone mill the typical annual production rate is in the range of 600,000–800,000 tons. With the continuous arrangement and depending on the product mix, this is increased to over 1.2 million tons and in many instances upwards of 1.5 million tons, which represents a significant increase. This is a major reason why almost all new installations in the early part of the twenty-first century are continuous mills, and why many of the older installations are being converted from stand-alone to continuous, so that the continuous mill is essentially replacing the stand-alone mill as the preferred arrangement.

To facilitate the conversion from stand-alone to continuous, strip turning units and bridles often are employed to match the tandem cold mill to the pickling process. The strip turning unit accommodates arrangements wherein the tandem cold rolling mill is not inline with (and almost always at right angles to) the pickling process by using a mechanical arrangement of cylinders and rollers to redirect the strip exiting the pickling line into the tandem cold mill. The bridle is an arrangement of several rolls powered by electric motors. The strip is passed well around each of the bridle rolls, with the bridle motors controlled to make up the difference in the strip tensions between the pickling process and the tandem cold mill.

Speed differences between the cold mill and the pickling process, during for example passage of the weld for the next coil in the cold mill, require a means to store the excess length of strip generated between the two processes due to the speed mismatch. The most commonly used device for this storage is the horizontal accumulator, wherein typically a car travels on a track which is a set of rails. The accumulator functions as a take-up device to store the excess strip, and then to reverse its direction of travel to deplete the excess strip from storage after the weld is through the mill and a new coil is started on the available rewind. Figure 1.5 shows a typical interface between a pickling process and a tandem cold mill using a bridle and an accumulator.

1.3.3 Main Drives and Motors

Modern tandem cold rolling mills are powered by variable speed AC machines which are synchronous motors or cage rotor induction motors. The drives which control these motors are mostly pulse-width-modulated (PWM) voltage source converters which are powered from a DC link with an active front end. This allows the transfer of energy between mill and reel drives that are motoring or regenerating, plus the potential for the transfer of energy back to the plant power system network. This particular drive topology injects very low harmonics into the power system so that distortion in voltage and current are reduced to very low levels without the need for harmonic filters. In addition, the active front end gives nearly

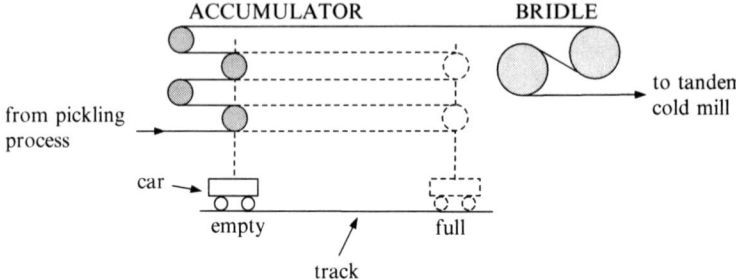

Fig. 1.5 Typical interface between pickling process and tandem cold mill

unity power without static VAR compensators. The efficiency of the drive itself is quite high, very nearly 99%. In some drive designs operation is maintained during certain network voltage drops without tripping. More detail regarding the current capabilities of motors and drives is provided in Chapter 6.

1.3.4 Subsequent Chapters

In Chapter 2 a model for the cold rolling process will be developed. Chapter 3 will present several basic conventional control approaches to the control of the various areas of the tandem cold rolling process. Chapter 4 will cover some advanced techniques and compare them to conventional methods. Chapter 5 deals specifically with an advanced technique based on the use of the state-dependent algebraic Riccati equation, and will compare this method with conventional methods based on performance and their usefulness in a practical setting. Chapter 6 covers some basic introductory material on motors and drives for powering modern tandem cold mills.

References

1. Bryant GF. The automation of tandem mills. London: The Iron and Steel Institute; 1973.
2. Carlton AJ, et al. Automation of the LTV Steel Hennepin tandem cold mill. Iron Steel Eng. 1992;69(6):17–28.
3. Duval P, Parks JC, Fellus G. Latest AGC technology installed at LTV's Cleveland 5-stand cold mill. Iron Steel Eng. 1991;69(11):46–51.
4. Geddes EJM, Postlewaite I. Improvements in product quality in tandem cold rolling using robust multivariable control. IEEE Trans Control Syst. Technol. 1998;6(2):257–69.
5. Hoshino I, et al. Observer-based multivariable control of the aluminum cold tandem mill. Automatica. 1988;24(6):741–54.
6. Roberts WL. The history of rolling. In: Cold rolling of steel. New York: Marcel Dekker; 1978.

Chapter 2
Process Model

2.1 Mathematical Model

A mathematical model of the tandem cold rolling process is a group of expressions which relate the rolling parameters to each other. Various mathematical models of the cold rolling process have been developed based on their intended use. For example, Roberts [1] identifies kinetic models which relate rolling force and spindle torque to other factors such as yield stress of the strip and strip tension, thermal models which include the aspects of kinetic models plus temperature distributions in the rolls and the strip, and economic models which are related to cost and profitability. The type of model desired for the work described in this book is one which relates the parameters of the tandem cold rolling process that are significant in the development of a process control strategy. In addition, the model must be useful in a practical sense, *i.e.* it must be capable of being implemented in a straightforward manner without being computationally demanding and yet retain the features needed for process control.

The cold rolling process involves the interaction of three components: (1) the work rolls, (2) the lubricant, and (3) the work piece (*i.e.* the strip). The roll force model, which predicts the deformation of the strip in the roll bite, is the center of the modeling of this process. Many existing classical roll force models are based on the theoretical work of Orowan [2], who used several simplifying assumptions to solve a differential equation developed by Von Karman [3] which expressed the pressure distribution across the arc of contact in the interface between the work rolls and the strip. However, these classical methods were computationally demanding and required considerable care to design numerical algorithms which were computationally robust. To attempt to provide a less complex roll force model which was more suitable for work involving control strategies, Bryant [4] developed simplified expressions for a model which reduced the problem to a series of algebraic equations. In addition, similar expressions for the prediction of neutral angle, slip, and torque were developed. Simulations were performed using the simplified expressions and the results were compared with the results obtained from simulations using the classical methods, which were extensively studied and verified. The comparisons generally showed close agreement between the simplified and classical simulations.

J. Pittner and M.A. Simaan, *Tandem Cold Metal Rolling Mill Control*,
Advances in Industrial Control, DOI 10.1007/978-0-85729-067-0_2,
© Springer-Verlag London Limited 2011

Considering this, and since data and simulation results are provided that are typical of practical applications with various products and operating conditions, the theory given in Bryant is used herein as a basis for development of a model that is somewhat simpler than Bryant's and than certain models developed by others (*e.g.* [5, 6]), but yet remains useful as a tool for the basic investigation of various control techniques. In addition to Bryant's work, certain empirical relationships given in Roberts [1] also are used in the development of the simplified model.

2.2 Theoretical System Equations

In this section theoretical system equations for relevant process variables are developed. The equations are simplified forms of the classical derivations which retain features relevant to the control of the tandem cold mill. The expressions given apply to each mill stand. Where adjacent mill stands are involved, the subscripts "*i*" and "*i* + *1*" are used, where *i* represents the mill stand number. Unless indicated otherwise, the subscript "*op*" indicates the desired value of a variable at an operating point, while the subscript "*0*" indicates the initial value of a variable.

2.2.1 Specific Roll Force

The theory for prediction of specific roll force is central to the development of a model for tandem cold rolling. In this section, theory is presented to provide some insight into phenomena occurring at the roll bite, and a simple but useful model is developed.

Referring to Figure 2.1, which approximately represents the strip in the roll bite area, the incoming strip is of thickness h_{in} at its centerline and is moving toward the roll bite with speed v_{in}. The strip exits the roll bite with thickness h_{out} at its centerline and with speed v_{out}. The incoming strip is under tension stress σ_{in}, the exiting strip is under tension stress σ_{out}.

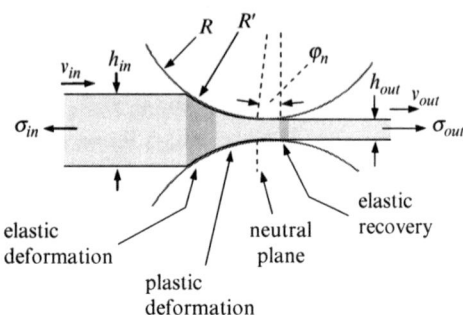

Fig. 2.1 Roll bite area

During rolling, the work roll is elastically flattened from its radius R to a radius R'. In the strip deformation process there is slipping between the flattened work roll surface and the strip, except at the neutral plane. At the entry to the neutral plane the strip speed is less than the peripheral speed of the work roll, and at the exit of the neutral plane the strip speed is greater than the peripheral speed of the work roll. At the neutral plane, the strip speed nearly matches the work roll peripheral speed, and the friction forces acting on the strip change sign. Three areas related to strip deformation are: (1) the zone of elastic deformation, (2) the zone of plastic deformation, and (3) the zone of elastic recovery.

Figure 2.2a shows an element of strip of unit width and length dx in the roll bite, between the entry of the plastic zone and the neutral plane, and at an angle φ with the deformed work roll radius R'. The change in thickness of the strip element is dh, and the compressive stresses exerted on the vertical walls of the element are σ and $(\sigma + d\sigma)$.

Figure 2.2b depicts the horizontal forces exerted on this element of strip in equilibrium. The radial pressure acting on the surface of the element is p_r, with a corresponding radial force $p_r dx/cos(\varphi)$, and horizontal components $p_r tan(\varphi)dx$. Friction force is $\mu p_r dx/cos(\varphi)$, with horizontal component $\mu p_r dx$, where μ is the coefficient of friction. Resolving horizontal forces on the element of strip in equilibrium gives

$$2p_r \sin(\phi)\frac{dx}{\cos(\phi)} - 2\mu p_r dx + \sigma h - (h + dh)(\sigma + d\sigma) = 0. \qquad (2.1)$$

By geometry,

$$\frac{dx}{\cos(\phi)} = R'd\phi, \qquad (2.2)$$

and substituting (2.2) into (2.1), grouping terms and neglecting $dh\, d\sigma$ gives

$$2P_r R'(\sin(\phi) \mp \mu\cos(\phi)) = d(h\sigma), \qquad (2.3)$$

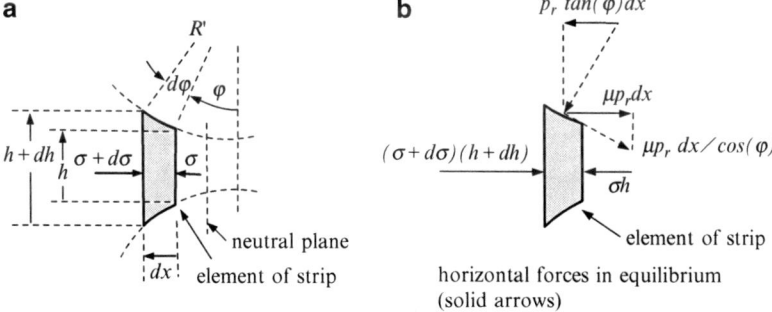

Fig. 2.2 Element of strip in the plastic zone

where symbols are as defined previously, and the plus sign applies for a strip element between the neutral plane and the exit of the plastic zone.

The Huber-Mises condition for plasticity [1] relates the roll pressure p_r to the principal stress σ in the horizontal plane and the compressive yield stress $k(h)$ as

$$p_r = \sigma + k(h), \tag{2.4}$$

where $k(h)$ denotes the dependency of k on h. Assuming this condition applies and substituting into (2.3) gives

$$\frac{d[h(p_r - k(h))]}{d\phi} = 2p_r R'(\sin(\phi) \mp \mu \cos(\phi)). \tag{2.5}$$

Using some assumptions and approximations, (2.5) can be solved to give a complicated expression for roll pressure p_r as a function of the angle φ in the plastic zone. An example of roll pressure variation in the roll bite area is depicted in Figure 2.3 which is a typical profile of roll pressure in the plastic and elastic zones as a function of the angle of the arc of contact over the entire roll bite.

The curve in Figure 2.3 is frequently denoted as a "friction hill" because of the shape of the upper portion, which results from the effects of the friction components. The peak of the curve is at the neutral plane where the friction forces change sign. The peak depends on the coefficient of friction (*i.e.* as the coefficient of friction is increased, the peak will increase and move toward the entry of the roll bite, and the specific roll force P will increase). The area under the curve is the specific roll force.

The deformed work roll radius is estimated by assuming an elliptical pressure distribution, using calculus, algebraic manipulations and some other assumptions.

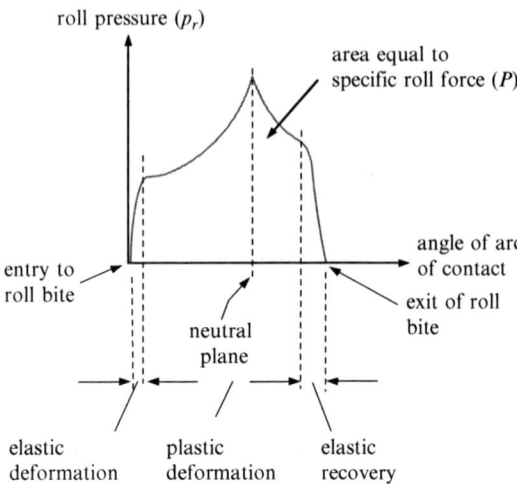

Fig. 2.3 Roll pressure vs angle in the roll bite

The resulting expression as developed by Hitchcock [1] for the deformed roll radius is

$$R' = R \left(1 + \frac{16(1 - v^2)P}{\pi E \left(h_{in} - h_{out} \right)} \right),$$
(2.6)

where v is Poisson's ratio, E is Young's modulus, and other symbols are as defined previously.

The compressive yield stress (or hardness) k of the strip affects the pressure distribution and thus the specific roll force P. Compressive yield stress is expressed as a base term k_0 plus an offset term $\Delta k_{\dot{e}}$ to account for dependency on strain rate. The following expression has been determined empirically [4], neglecting thermal effects

$$k = k_0 + \Delta k_{\dot{e}},$$
(2.7)

where

$$k_0 = a(b + r)^c,$$
(2.8)

a, b, and c are constants, r is the reduction corresponding to a thickness h with respect to an annealed thickness H_a, i.e.

$$r = \frac{H_a - h}{H_a},$$
(2.9)

and $\Delta k_{\dot{e}}$ is a speed dependent offset due to strain rate, which is approximated [1] by

$$\Delta k_{\dot{e}} = \gamma \left(3 + \log_{10} \frac{V}{h_{in0}} \sqrt{\frac{h_{in0} - h_{out0}}{R}} \right),$$
(2.10)

where V is work roll peripheral speed, h_{in0} and h_{out0} are the desired initial values at an operating point, and γ is a constant. The expression for r accounts for work hardening effects caused by rolling in upstream mill stands, and any processing after hot rolling and before entry into the tandem cold mill. The annealed thickness H_a is the strip thickness at the exit of the hot rolling process. H_a is sometimes significantly different than the thickness at the tandem cold mill entry because of processing after hot rolling and prior to cold rolling. The interstand time delay approximation (2.28) is used to estimate the delay in changes in the compressive yield stress from stand i to stand $i + 1$.

The coefficient of friction μ, which is speed dependent, also affects the pressure distribution. Empirically, the coefficient of friction is determined [1] as

$$\mu = \sqrt{\left(\frac{h_{in0} - h_{out0}}{2R} \right)} \left(0.5 + (K_1 - 0.5) e^{-K_2 V} \right),$$
(2.11)

where K_1 and K_2 are constants.

The roll pressure distribution depicted in Figure 2.3 can be approximated by graphical methods which result in the following simplified expression for specific roll force in the plastic zone, where specific roll force in the elastic zones is assumed negligible for most practical cases,

$$P = (\bar{k} - \bar{\sigma})\sqrt{R'\delta}\,(1 + 0.4\alpha). \tag{2.12}$$

In (2.12) $\bar{k}$ is the compressive yield stress given by

$$\bar{k} = \lambda_1 k_{in} + (1 - \lambda_1)k_{out} \tag{2.13}$$

where k_{in} is the compressive yield stress at the entry of the plastic zone, k_{out} is the compressive yield stress at the exit of the plastic zone, and λ_1 is a constant and $\bar{\sigma}$ is the mean tension stress given by

$$\bar{\sigma} = \lambda_2 \sigma_{in} + (1 - \lambda_2)\,\sigma_{out}, \tag{2.14}$$

here λ_2 is a constant, α is given by

$$\alpha = \sqrt{\frac{h_{out}}{h_{in}}}\exp\left(\frac{\mu\sqrt{R'\delta}}{\bar{h}}\right) - 1, \tag{2.15}$$

δ is the draft (*i.e.*, $h_{out} - h_{in}$), $\bar{h}$ is the mean thickness given by

$$\bar{h} = \beta\,h_{out} + (1 - \beta)h_{in}, \tag{2.16}$$

and β is a constant. Other symbols in (2.12)–(2.15) are as defined previously.

2.2.2   Work Roll Torque

A simple approximation of work roll torque is obtained by assuming a constant pressure over the arc of contact, and taking the total force exerted on the strip as the draft (*i.e.*, $h_{in} - h_{out}$) multiplied by the average yield stress, with zero tension stress at the entry and exit ends of the roll bite,

$$G = R\,(h_{in} - h_{out})\,\bar{\sigma}, \tag{2.17}$$

where G is the specific total work roll torque, and other symbols are as defined before. With tension on each side of the roll bite

$$G = R\bar{\sigma}\left[(h_{in} - h_{out})\left(1 + \frac{\sigma_{out}}{\bar{\sigma}}\right) + \left(\frac{\sigma_{in} - \sigma_{out}}{\bar{\sigma}}\right)\right]. \tag{2.18}$$

Since work roll speed is determined by closed-loop speed control, excursions in work roll speed caused by changes in torque can be taken as less significant so that (2.17) and (2.18) are provided only as background information, and need not be included in the basic model for the purposes of the work considered herein. Additional efforts beyond the scope of this work may expand the model given in Sect. 2.2.9 for work roll speed and consider the effects of work roll torque.

2.2.3 Forward Slip

The forward slip f is given approximately by

$$f = \frac{V_{out} - V}{V},\tag{2.19}$$

where V_{out} is the strip speed at the exit of the roll bite, and V is the peripheral speed of the work roll. By conservation of volume through the roll bite and assuming constant strip width,

$$f = \frac{h_n - h_{out}}{h_{out}},\tag{2.20}$$

where h_n is the thickness at the neutral plane. The angle at the neutral plane is approximated [4] by

$$\phi_n = \frac{1}{2}\frac{h_{out}}{\tilde{h}}\sqrt{\frac{\delta}{R'}} - \frac{1}{4}\frac{h_{out}\delta}{\tilde{h}\mu R'} + \frac{1}{4}\frac{h_{out}}{\mu R'}\left[\frac{\sigma_{out}}{k_{out}} - \frac{\sigma_{in}}{k_{in}}\right],\tag{2.21}$$

where $\tilde{h}$ is the mean thickness (2.16) except with the value of the constant β adjusted slightly. The angle of contact is approximated by

$$\phi_1 = \sqrt{\left(\frac{h_{in} - h_{out}}{R'}\right)}.\tag{2.22}$$

Analysis using an element of strip just at the exit side of the neutral plane and using (2.19), (2.20), and (2.22) gives an expression for the forward slip that is useful for control development

$$f = \left(\frac{h_{in} - h_{out}}{h_{out}}\right)\left(\frac{\phi_n}{\phi_1}\right)^2,\tag{2.23}$$

where the symbols are as defined previously.

2.2.4   Interstand Tension Stress

An equation for interstand tension stress is developed by applying Hooke's law to a length of strip between successive mill stands, assuming some average strip thickness and neglecting stretching of the strip. The average tension stress is

$$\sigma_{i,i+1} = E\frac{\Delta l}{L_0},\tag{2.24}$$

where L_0 is the distance between mill stands i and $i+1$, Δl is the change in length due to application of a tension force corresponding to $\sigma_{i,i+1}$ and E is Young's modulus. Over an increment of time dt

$$\frac{d(\sigma_{i,i+1})}{dt} \equiv \dot\sigma_{i,i+1} = \frac{E(V_{in,i+1} - V_{out,i})}{L_0},\ \sigma_{i,i+1}(0) = \sigma_{0,i,i+1},\tag{2.25}$$

where $V_{in,i+1}$ is the strip velocity at the input to stand $i+1$, $V_{out,i}$ is the strip velocity at the output of stand i, and $\sigma_{0,i,i+1}$ is the initial tension stress in the strip between stands i and $i+1$.

2.2.5   Output Thickness

A linear approximation for output thickness is

$$h_{out} = S + S_0 + \frac{PW}{M},\tag{2.26}$$

where P is the specific roll force, W is the strip width, S is work roll actuator position, S_0 is the intercept of the linearized portion of the mill stretch curve, and M is the mill modulus. More detail related to the usage of this approximation is presented in Chapter 3.

2.2.6   Interstand Time Delay

The interstand time delay $\tau_{d,i,i+1}$ is the time taken for a small element of strip to travel a distance L_0 from the exit of stand i to the entry of stand $i+1$, and is defined implicitly as

$$L_0 = \int_0^{\tau_{d,i,i+1}} V_{out,i}(t)dt.\tag{2.27}$$

At any instant of time the time delay is approximated as

$$\tau_{d,i,i+1} = \frac{L_0}{V_{out,i}}.$$
(2.28)

2.2.7 Input Thickness

The input thickness to stand 1 is the input thickness to the mill. The input thickness to stands 2,3,4, and 5 is the output thickness from the previous stand delayed by the interstand time delay (2.28).

2.2.8 Work Roll Actuator Position

The work roll actuator position controller is modeled (closed-loop) as a single first order lag, based on typical data. The model is

$$\dot{S} = \frac{U_S}{\tau_S} - \frac{S}{\tau_S}, \; S(0) = S_0,$$
(2.29)

where S is the work roll actuator position, U_S is the position controller reference, τ_S is the closed-loop time constant, and S_0 is the initial work roll actuator position.

2.2.9 Work Roll Speed

The work roll drive speed controller is modeled (closed-loop) as a single first order lag, based on application experience. The model is

$$\dot{V} = \frac{U_V}{\tau_V} - \frac{V}{\tau_V}, \; V(0) = V_0,$$
(2.30)

where V is the work roll peripheral speed, U_V is the drive speed reference, τ_V is the closed-loop time constant, and V_0 is the initial work roll peripheral speed.

2.3 State and Output Equations

In this section, the relationships developed in Sect. 2.2 are put into the forms of a state equation (2.31) and an output equation (2.32),

$$\frac{dx}{dt} \equiv \dot{x} = a(x) + Bu, \; x(0) = x_0$$
(2.31)

$$y = g(x),$$
(2.32)

where $x \in R^n$ is a vector whose elements represent the individual state variables, $a(x)$ is a state-dependent vector, $u \in R^m$ is a vector whose elements represent the individual control variables, $y \in R^p$ is a vector whose elements represent the individual output variables, $g(x) \in R^p$ is a state-dependent vector, and $B \in R^{nxm}$ is a constant matrix. The individual state variables, control variables, and output variables represented by the elements of the vectors x, u, and y respectively in (2.31) and (2.32) are as shown in Table 2.1, where the variables and subscripts are as defined previously.

2.4   Model Verification

The model was verified by simulation using three operating points with reduction patterns similar to the typical production schedules given by Bryant [4]. Since it is impractical to use a physical mill to perform open-loop tests to verify the model, the results are compared to the results of Byrant's simulations. Bryant's results have been verified by comparison against results of simulations using classical approaches and against data derived from actual rolling operations, and are considered by control theoreticians and practitioners as an industry "benchmark" for verification of models of the tandem cold rolling process. Additionally, the simulation results are compared against the results of Geddes [5] who has developed models which are based on reduction patterns similar to Bryant's. In addition, certain results are compared with the results of Guo [6] whose model is based on reduction patterns that also are somewhat similar in several areas to Bryant's.

The simulations were open-loop (*i.e.* the mill is uncontrolled), with mill exit tension assumed to be held constant by the coiler controller. To be consistent with Bryant, mill entry tension is assumed to be about zero which is somewhat typical for an older mill with a coil box at the entry end. However, assuming an entry coiler where entry tension is controlled to be other than zero does not significantly affect the results, as was shown by additional simulations. Small step changes, both positive and negative, in each of the following variables were applied separately, with the other listed variables remaining at their operating point values.

Table 2.1 Variable assignments for state, control, and output vectors

State vector		Control vector		Output vector	
$x_1 (\sigma_{12})$	$x_8 (S_4)$	$u_1 (U_{S1})$	$u_6 (U_{V1})$	$y_1 (h_{out1})$	$y_8 (\sigma_{34})$
$x_2 (\sigma_{23})$	$x_9 (S_5)$	$u_2 (U_{S2})$	$u_7 (U_{V2})$	$y_2 (h_{out2})$	$y_9 (\sigma_{45})$
$x_3 (\sigma_{34})$	$x_{10} (V_1)$	$u_3 (U_{S3})$	$u_8 (U_{V3})$	$y_3 (h_{out3})$	$y_{10} (P_1)$
$x_4 (\sigma_{45})$	$x_{11} (V_2)$	$u_4 (U_{S4})$	$u_9 (U_{V4})$	$y_4 (h_{out4})$	$y_{11} (P_2)$
$x_5 (S_1)$	$x_{12} (V_3)$	$u_5 (U_{S5})$	$u_{10} (U_{V5})$	$y_5 (h_{out5})$	$y_{12} (P_3)$
$x_6 (S_2)$	$x_{13} (V_4)$			$y_6 (\sigma_{12})$	$y_{13} (P_4)$
$x_7 (S_3)$	$x_{14} (V_5)$			$y_7 (\sigma_{23})$	$y_{14} (P_5)$

The initial simulations were performed at run speed, which was taken to be the maximum speed of the mill, and it was assumed that the mill is designed to successfully process the product being rolled at this speed. The variables which were perturbed are:

- Strip annealed thickness
- Strip thickness at the entry to stand 1, accompanied by an equal change in strip annealed thickness
- Strip hardness (compressive yield stress) at the entry to stand 1
- Coefficient of friction at stands 1, 3, and 5 (individually)
- Work roll actuator position references of stands 1, 3, and 5 (individually) and
- Work roll peripheral speed references of stands 1, 3, and 5 (individually)

The mill speed was lowered to thread speed (5% of run speed) and the above repeated. Steady-state and dynamic responses to selected perturbations were recorded and compared to results presented in Bryant and to the results presented in Geddes for the application of the same perturbations in both cases.

In the above it should be noted that the strip annealed thickness is the strip thickness as it leaves the hot rolling process, and that the strip thickness at the entry to the tandem cold rolling process is not always the same as the strip annealed thickness. This is because changes in strip thickness sometimes occur during processing that is performed between the exit of the hot rolling process and the entry of the tandem cold rolling process.

2.4.1 Operating Points

Table 2.2 presents three production schedules taken from Bryant and used for model verification, and also used by Geddes for model verification. Table 2.3 presents the mill and strip parameters used. Prior to the application of perturbations to the mill model, three steady-state operating points (Table 2.4) were established using the three production schedules of Table 2.2. The following procedure was used to set the three operating points, assuming an uncontrolled mill:

- The strip speed at the mill exit was set to 1,220 m/min which was taken as the maximum mill speed, *i.e.* 100% speed, or run speed
- The individual work roll peripheral speeds were set based on estimations using the values of forward slips given in Bryant
- The work roll actuator positions were adjusted to set the specific roll forces to closely approximate the reduction patterns of Table 2.2
- Data were recorded for the exit thickness and the total roll force for each mill stand, and the interstand tensions and
- Comparisons were made with the results of Bryant and Geddes who both used the same production schedules

Table 2.2 Production schedules

Parameter	Production schedule number		
	1	2	3
Mill entry thickness, mm	3.56	2.36	1.78
Exit thickness, stand 1	2.95	2.01	1.22
Exit thickness, stand 2	2.44	1.52	0.79
Exit thickness, stand 3	2.01	1.22	0.56
Exit thickness, stand 4	1.68	0.97	0.38
Exit thickness, stand 5	1.58	0.91	0.36
Tension stress, mill entry, kN/mm^2	0.0	0.0	0.0
Tension stress, stands 1,2	0.086	0.091	0.097
Tension stress, stands 2,3	0.088	0.094	0.105
Tension stress, stands 3,4	0.089	0.097	0.119
Tension stress, stands 4,5	0.092	0.102	0.138
Tension stress, mill exit	0.028	0.028	0.028

Table 2.3 Mill and strip parameters

Parameter	Dimension
Work roll radius	292 mm
Mill modulus	3,921 kN/mm
Distance between stands	4,318 mm
Strip width	914 mm
Annealed thickness/mill entry thickness	1.095
Young's modulus	207 kN/mm^2
Poisson's ratio	0.3

Table 2.4 Operating points

Parameter	Operating point number		
	1	2	3
Mill entry thickness, mm	3.56	2.36	1.78
Exit thickness, stand 1	2.95	2.01	1.22
Exit thickness, stand 2	2.44	1.52	0.79
Exit thickness, stand 3	2.01	1.22	0.56
Exit thickness, stand 4	1.68	0.97	0.38
Exit thickness, stand 5	1.58	0.91	0.36
Tension stress, mill entry, kN/mm^2	0.0	0.0	0.0
Tension stress, stands 1,2	0.080	0.103	0.111
Tension stress, stands 2,3	0.078	0.126	0.132
Tension stress, stands 3,4	0.057	0.096	0.132
Tension stress, stands 4,5	0.055	0.060	0.085
Tension stress, mill exit	0.028	0.028	0.028

2.4.2 Simulation Results

Step changes in the variables noted previously were applied to the model for each of the three operating points at the maximum mill exit speed (100%, 1,220 m/min).

For the purposes of comparison to the results of Bryant and to the results of Geddes, the sizes of the step changes were approximately those given in Bryant, which also were used by Geddes. Table 2.5 summarizes the mean and standard deviation of the percent deviations from the values of Bryant and Geddes using the data of Tables 2.6–2.17, which present the results for each of the cases given in Bryant. In these tables the percent change in a variable represents an average of the results for the three operating points. These data provide a thorough comparison with the "benchmark" results of Bryant and give some insight into the open-loop character-istics of the tandem mill for the thread and run regimes of mill operation.

The strip speed at the mill exit was lowered to thread speed (61 m/min, *i.e.* 5% of top speed) and the simulations repeated. Steady-state responses considered more pertinent to the development of the control strategy are presented in Tables 2.18–2.25. Steady-state responses at 100% speed are also shown to facilitate comparison.

Negative step changes of the same magnitude as the positive step changes were applied at 100% speed and at 5% speed. In general the resulting magnitudes of the output thickness and roll force did not change significantly from the magnitudes resulting from application of the positive steps. Some changes in the magnitudes of interstand tension occurred in those instances where the change in interstand tension was not large. Where the change in interstand tension was larger (*e.g.* a step change in hardness), the change in magnitude was of the same order as the change in magnitude with the positive step. Some typical results are depicted in Tables 2.26–2.29 for 100% speed. The results for 5% speed are similar. The percent changes for positive step changes also are shown to facilitate comparison.

Table 2.5 Variations of model from Bryant and Geddes

Variable	Mean of percent deviations of model from		Standard deviation of percent deviations of model from	
	Bryant	Geddes	Bryant	Geddes
Output thickness	−4.5	−2.7	25.7	13.1
Total roll force	2.0	−23.6	40.6	23.5
Interstand tension	3.8	68.7	73.7	203.3

Table 2.6 +2% Step change in stand 1 input thickness

Variable	Source	Percent change in variable (steady-state)				
		Stand 1	Stand 2	Stand 3	Stand 4	Stand 5
Output thickness	Bryant	1.8	1.8	1.8	1.8	1.7
	Geddes	2.32	2.13	1.90	2.14	2.21
	Model	2.40	2.38	2.39	2.31	2.42
Total roll force	Bryant	2.0	1.3	0.9	0.5	0.8
	Geddes	3.23	2.27	1.30	1.23	1.68
	Model	2.17	1.67	1.35	1.09	1.57
Interstand tension	Bryant	−trace	−trace	+trace	−trace	–
	Geddes	+trace	1.7	4.6	−2.4	–
	Model	1.2	0.7	0.4	8.8	–

Table 2.7 +5% Step change in annealed thickness

Variable	Source	Percent change in variable (steady-state)				
		Stand 1	Stand 2	Stand 3	Stand 4	Stand 5
Output thickness	Bryant	2.5	2.6	2.6	2.7	2.6
	Model	1.42	1.52	1.51	1.46	1.53
Total roll force	Bryant	2.8	2.4	1.9	0.8	1.7
	Model	1.27	1.13	1.00	0.91	1.11
Interstand tension	Bryant	6.0	5.0	4.7	2.3	–
	Model	9.9	6.9	6.2	10.1	–

Table 2.8 +5% Step change in stand 1 input hardness

Variable	Source	Percent change in variable (steady-state)				
		Stand 1	Stand 2	Stand 3	Stand 4	Stand 5
Output thickness	Bryant	2.1	3.0	3.3	3.4	3.7
	Geddes	1.23	1.47	1.66	1.49	1.52
	Model	1.27	1.51	1.56	1.56	1.86
Total roll force	Bryant	2.3	2.4	2.1	1.4	3.1
	Geddes	1.79	1.61	1.24	0.92	1.30
	Model	1.07	1.00	0.84	0.71	1.17
Interstand tension	Bryant	24.7	42.7	49.0	54.0	–
	Geddes	7.2	5.7	6.7	20.4	–
	Model	26.1	31.9	43.7	61.4	–

Table 2.9 +10% Step change in stand 1 friction coefficient

Variable	Source	Percent change in variable (steady-state)				
		Stand 1	Stand 2	Stand 3	Stand 4	Stand 5
Output thickness	Bryant	0.4	0.4	0.4	0.4	0.30
	Geddes	0.61	0.54	0.48	0.55	0.50
	Model	0.98	0.95	0.95	0.93	0.96
Total roll force	Bryant	+trace	+trace	+trace	+trace	+trace
	Geddes	0.70	0.40	0.21	0.19	0.24
	Model	0.63	0.43	0.37	0.30	0.43
Interstand tension	Bryant	−2.0	−trace	−trace	−trace	–
	Geddes	−2.9	−2.7	2.2	7.1	–
	Model	−0.8	−0.4	−0.4	2.0	–

Table 2.10 +10% Step change in stand 3 friction coefficient

Variable	Source	Percent change in variable (steady-state)				
		Stand 1	Stand 2	Stand 3	Stand 4	Stand 5
Output thickness	Bryant	+trace	−0.2	+trace	0	0
	Geddes	+trace	−0.26	+trace	+trace	+trace
	Model	0.05	−0.24	0	0	0
Total roll force	Bryant	+trace	−trace	0	0	0
	Geddes	+trace	0.26	0	0	−trace
	Model	0	−0.16	0	0	0
Interstand tension	Bryant	0	10.0	0	0	–
	Geddes	−2.8	3.0	−1.1	9.4	–
	Model	−1.4	13.8	−0.8	−0.4	–

Table 2.11 +10% Step change in stand 5 friction coefficient

Variable	Source	Percent change in variable (steady-state)				
		Stand 1	Stand 2	Stand 3	Stand 4	Stand 5
Output thickness	Bryant	0	0	+trace	−trace	+trace
	Model	0	0	0	−trace	0
Total roll force	Bryant	0	0	0	−trace	0
	Model	0	0	0	0	0
Interstand tension	Bryant	0	0	−trace	7.3	−
	Model	0	+trace	−trace	5.9	−

Table 2.12 +0.1 mm Step change in stand 1 position actuator reference

Variable	Source	Percent change in variable (steady-state)				
		Stand 1	Stand 2	Stand 3	Stand 4	Stand 5
Output thickness	Bryant	1.5	1.3	1.3	1.3	1.2
	Geddes	2.42	2.04	1.77	2.10	2.04
	Model	1.55	1.42	1.41	1.37	1.42
Total roll force	Bryant	−3.5	0.8	0.3	0.3	0.8
	Geddes	−4.42	2.17	1.17	1.22	1.64
	Model	−3.23	0.95	0.76	0.62	0.88
Interstand tension	Bryant	−6.3	−5.3	−1.8	−1.8	−
	Geddes	−3.5	9.7	2.1	−0.6	−
	Model	−0.5	−1.4	−1.5	2.5	−

Table 2.13 +0.1 mm Step change in stand 3 position actuator reference

Variable	Source	Percent change in variable (steady-state)				
		Stand 1	Stand 2	Stand 3	Stand 4	Stand 5
Output thickness	Bryant	+trace	−0.4	0.4	+trace	+trace
	Geddes	0.16	−0.43	0.55	0	0
	Model	+trace	−0.27	0.34	0	0
Total roll force	Bryant	0	−trace	−4.3	0	0
	Geddes	0.14	−0.92	−6.17	+trace	+trace
	Model	0	−0.30	−4.60	0	0
Interstand tension	Bryant	−trace	18.7	1.7	0	−
	Geddes	−1.9	9.7	2.1	−trace	−
	Model	−1.5	23.1	6.8	2.1	−

Table 2.14 +0.1 mm Step change in stand 5 position actuator reference

Variable	Source	Percent change in variable (steady-state)				
		Stand 1	Stand 2	Stand 3	Stand 4	Stand 5
Output thickness	Bryant	0	0	+trace	−0.5	0.2
	Model	0	0	0	−0.38	0.23
Total roll force	Bryant	0	0	0	0	−7.8
	Model	0	0	0	−trace	−7.39
Interstand tension	Bryant	0	0	−1.8	24.0	−
	Model	−trace	0.2	−1.1	37.4	−

Table 2.15 +2% Step change in stand 1 speed actuator reference

Variable	Source	Percent change in variable (steady-state)				
		Stand 1	Stand 2	Stand 3	Stand 4	Stand 5
Output thickness	Bryant	0.5	1.6	1.8	1.9	1.8
	Geddes	0.44	1.47	1.47	1.86	1.78
	Model	0.46	1.57	1.81	1.82	1.85
Total roll force	Bryant	+trace	1.6	1.4	0.8	1.3
	Geddes	0.66	1.67	1.09	1.24	1.64
	Model	0.35	1.27	1.16	1.01	1.30
Interstand tension	Bryant	−18.3	−8.0	−4.7	−5.3	−
	Geddes	−10.0	−5.1	−0.2	−10.4	−
	Model	−27.0	−6.9	−3.6	3.3	−

Table 2.16 +2% Step change in stand 3 speed actuator reference

Variable	Source	Percent change in variable (steady-state)				
		Stand 1	Stand 2	Stand 3	Stand 4	Stand 5
Output thickness	Bryant	0	−0.5	−1.4	−0.2	0
	Geddes	+trace	−0.52	−1.07	+trace	0
	Model	+trace	−0.53	−1.30	−0.13	−trace
Total roll force	Bryant	0	−trace	−trace	0	0
	Geddes	+trace	−0.66	−0.83	0	−trace
	Model	0	−0.42	−0.72	−trace	−trace
Interstand tension	Bryant	0	20.0	−13.0	−6.0	−
	Geddes	−1.1	10.0	−5.2	0.3	−
	Model	−1.2	29.9	−30.5	−14.5	−

Table 2.17 +2% Step change in stand 5 speed actuator reference

Variable	Source	Percent change in variable (steady-state)				
		Stand 1	Stand 2	Stand 3	Stand 4	Stand 5
Output thickness	Bryant	0	0	+trace	−0.8	−1.9
	Model	0	0	+trace	−1.01	−1.61
Total roll force	Bryant	0	0	0	−trace	−trace
	Model	0	0	0	−0.5	−1.10
Interstand tension	Bryant	0	0	−7.0	46.7	−
	Model	−0.1	0.5	−1.4	97.6	−

Dynamic responses of stand exit thicknesses, interstand tensions, and stand specific roll forces to step changes in the variables noted previously were taken with operation at 100% speed and at 5% speed using operating point 1 (Table 2.4). In all cases the model was stable and there were no undesirable excursions in the responses. Figures 2.4–2.6 depict dynamic responses for three selected cases at 100% speed:

- +2% step change in the input thickness to stand 1
- +2% step change in the stand 1 position actuator reference and
- +2% step change in the stand 1 speed actuator reference

Table 2.18 +2% Step change in stand 1 input thickness at 5% speed

Variable	Speed (%)	Percent change in variable (steady-state)				
		Stand 1	Stand 2	Stand 3	Stand 4	Stand 5
Output thickness	100	2.40	2.38	2.39	2.31	2.42
	5	2.39	2.35	2.33	2.27	2.40
Total roll force	100	2.17	1.67	1.35	1.09	1.57
	5	2.16	1.65	1.34	1.09	1.56
Interstand tension	100	1.2	0.7	0.4	8.8	–
	5	0.6	+trace	−0.9	9.2	–

Table 2.19 +5% Step change in stand 1 input hardness at 5% speed

Variable	Speed (%)	Percent change in variable (steady-state)				
		Stand 1	Stand 2	Stand 3	Stand 4	Stand 5
Output thickness	100	1.27	1.51	1.56	1.56	1.86
	5	1.23	1.31	1.41	1.44	1.64
Total roll force	100	1.07	1.00	0.84	0.71	1.17
	5	1.00	0.91	0.75	0.62	1.05
Interstand tension	100	26.1	31.9	43.7	61.4	–
	5	20.3	23.0	24.9	55.5	–

Table 2.20 +0.1 mm Step change in stand 1 position actuator ref at 5% speed

Variable	Speed (%)	Percent change in variable (steady-state)				
		Stand 1	Stand 2	Stand 3	Stand 4	Stand 5
Output thickness	100	1.55	1.42	1.41	1.37	1.42
	5	1.41	1.21	1.21	1.18	1.25
Total roll force	100	−3.23	0.95	0.76	0.62	0.88
	5	−3.29	0.87	0.70	0.57	0.81
Interstand tension	100	−0.5	−1.4	−1.5	2.5	–
	5	−0.1	−1.1	−1.5	3.0	–

Table 2.21 +0.1 mm Step change in stand 3 position actuator ref at 5% speed

Variable	Speed (%)	Percent change in variable (steady-state)				
		Stand 1	Stand 2	Stand 3	Stand 4	Stand 5
Output thickness	100	+trace	−0.27	0.34	0	0
	5	0	−0.16	0.38	0	0
Total roll force	100	0	−0.30	−4.60	0	0
	5	0	−0.23	−4.61	0	0
Interstand tension	100	−1.5	23.1	6.8	2.1	–
	5	−1.4	15.8	4.2	2.1	–

2.5 Concluding Comments

The reduction patterns of the three operating points (Table 2.4) closely match those of the three production schedules (Table 2.2), however there are some discrepancies in the interstand tensions. As previously noted, these discrepancies do not result in

Table 2.22 +0.1 mm Step change in stand 5 position actuator ref at 5% speed

Variable	Speed (%)	Percent change in variable (steady-state)				
		Stand 1	Stand 2	Stand 3	Stand 4	Stand 5
Output thickness	100	0	0	0	−0.38	0.23
	5	0	0	0	−0.19	0.43
Total roll force	100	0	0	0	−trace	−7.39
	5	0	0	0	−0.13	−7.35
Interstand tension	100	−trace	0.2	−1.1	37.4	–
	5	−trace	0.3	−1.7	38.0	–

Table 2.23 +2% Step change in stand 1 speed actuator ref at 5% speed

Variable	Speed (%)	Percent change in variable (steady-state)				
		Stand 1	Stand 2	Stand 3	Stand 4	Stand 5
Output thickness	100	0.46	1.57	1.81	1.82	1.85
	5	0.56	1.72	1.98	1.82	2.17
Total roll force	100	0.35	1.27	1.16	1.01	1.30
	5	0.50	1.23	1.12	1.59	1.32
Interstand tension	100	−27.0	−6.9	−3.6	3.3	–
	5	−23.0	−5.0	−3.0	4.8	–

Table 2.24 +2% Step change in stand 3 speed actuator ref at 5% speed

Variable	Speed (%)	Percent change in variable (steady-state)				
		Stand 1	Stand 2	Stand 3	Stand 4	Stand 5
Output thickness	100	+trace	−0.53	−1.30	−0.13	−trace
	5	+trace	−0.64	−1.30	−0.10	0
Total roll force	100	0	−0.42	−0.72	−trace	−trace
	5	0	−0.43	−0.76	−trace	0
Interstand tension	100	−1.2	29.9	−30.5	−14.5	–
	5	−1.8	24.8	−21.6	−10.5	–

Table 2.25 +2% Step change in stand 5 speed actuator ref at 5% speed

Variable	Speed (%)	Percent change in variable (steady-state)				
		Stand 1	Stand 2	Stand 3	Stand 4	Stand 5
Output thickness	100	0	0	+trace	−1.01	−1.61
	5	0	0	0.15	−0.78	−1.50
Total roll force	100	0	0	0	−0.50	−1.10
	5	0	0	0	−0.39	−1.07
Interstand tension	100	−0.1	0.5	−1.4	97.6	–
	5	−0.1	0.7	−3.9	102.6	–

significant increases or decreases in interstand tensions which would cause the strip to break or become slack. These discrepancies might be attributed to uncertainties in the estimation of forward slips under open-loop conditions. Additionally, the

Table 2.26   2% Step change in stand 1 input thickness

Variable	Step	Percent change in variable (steady-state)				
		Stand 1	Stand 2	Stand 3	Stand 4	Stand 5
Output thickness	Pos	2.40	2.38	2.39	2.31	2.42
	Neg	−2.48	−2.45	−2.43	−2.55	−2.35
Total roll force	Pos	2.17	1.67	1.35	1.09	1.57
	Neg	−2.22	−1.69	−1.36	−1.19	−1.51
Interstand tension	Pos	1.2	0.7	0.4	8.8	−
	Neg	−1.0	0.1	0.1	10.2	−

Table 2.27   −5% Step change in stand 1 input hardness

Variable	Step	Percent change in variable (steady-state)				
		Stand 1	Stand 2	Stand 3	Stand 4	Stand 5
Output thickness	Pos	1.27	1.51	1.56	1.56	1.86
	Neg	−1.34	−1.60	−1.62	−1.81	−1.96
Total roll force	Pos	1.07	1.00	0.84	0.71	1.17
	Neg	−1.13	−1.05	−0.88	−0.82	−1.25
Interstand tension	Pos	26.1	31.9	43.7	61.4	−
	Neg	−25.7	−31.4	−43.5	−44.5	−

Table 2.28   −0.1 mm Step change in stand 1 position actuator reference

Variable	Step	Percent change in variable (steady-state)				
		Stand 1	Stand 2	Stand 3	Stand 4	Stand 5
Output thickness	Pos	1.55	1.42	1.41	1.37	1.42
	Neg	−1.55	−1.43	−1.41	−1.47	−1.39
Total roll force	Pos	−3.23	0.95	0.76	0.62	0.88
	Neg	−3.21	−0.96	−0.77	−0.65	−0.86
Interstand tension	Pos	−0.5	−1.4	−1.5	2.5	−
	Neg	1.6	1.7	3.1	0.0	−

Table 2.29   −2% Step change in stand 1 speed actuator reference

Variable	Step	Percent change in variable (steady-state)				
		Stand 1	Stand 2	Stand 3	Stand 4	Stand 5
Output thickness	Pos	0.46	1.57	1.81	1.82	1.85
	Neg	−0.49	−1.68	−1.85	−1.95	−1.85
Total roll force	Pos	0.35	1.27	1.16	1.01	1.30
	Neg	−0.44	−1.33	−1.02	−0.91	−1.18
Interstand tension	Pos	−27.0	−6.9	−3.6	3.3	−
	Neg	26.6	7.1	3.8	7.3	−

differences in annealed thicknesses, friction coefficients, hardness functions, or a combination of these, all of which are unavailable in [4] specifically for the three schedules being considered may further contribute to the discrepancies. Geddes' results also display discrepancies from [4] in interstand tensions and Geddes

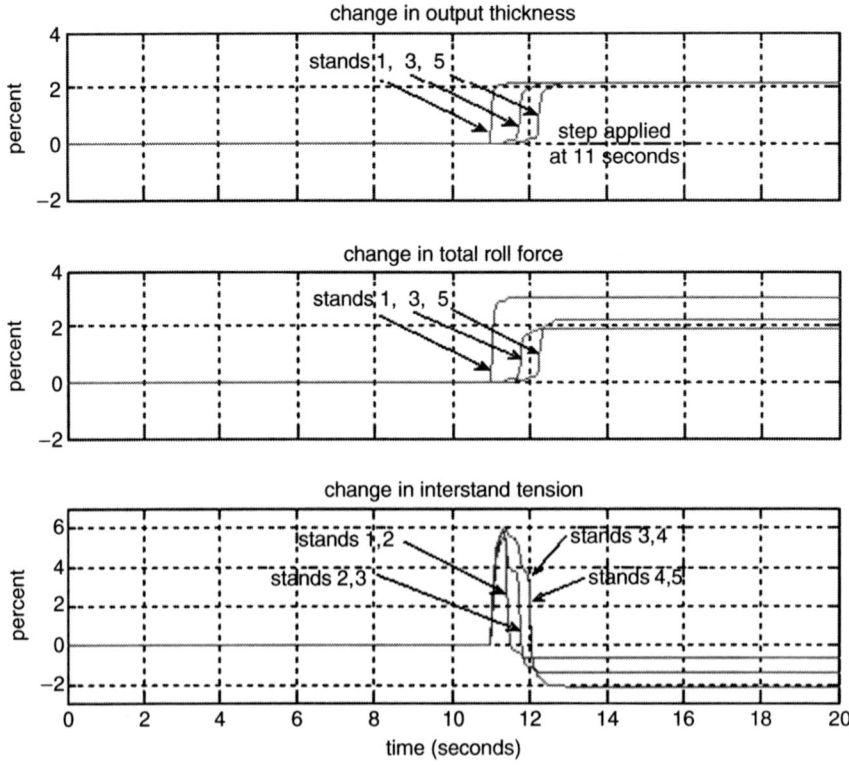

Fig. 2.4 Responses to +2% step change in input thickness

attributes them mostly to differences in the estimates of the friction coefficients. However, under closed-loop control as described in Chapter 5 the interstand tensions are determined as a result of control action and are consistent with the values given in the production schedules, with the forward slips adjusted inherently by the process controller.

The application of step changes in input thickness, annealed thickness, and input hardness caused changes in the steady-state magnitudes of output thicknesses which differ somewhat from those noted in [4, 5]. These differences are not severe and may be attributed to slight differences in modeling, differences in certain operating point values, or both, and are considered acceptable for the intended purposes of the model. More significantly, the directions of the changes in the steady-state output thicknesses are in good agreement with [4, 5], as are the directions in the changes in roll forces and the directions of the larger changes in interstand tensions.

The application of step changes in actuator references produced steady-state changes in output thicknesses and total roll forces that closely matched [4] both in magnitude and direction. Steady-state changes in interstand tensions generally

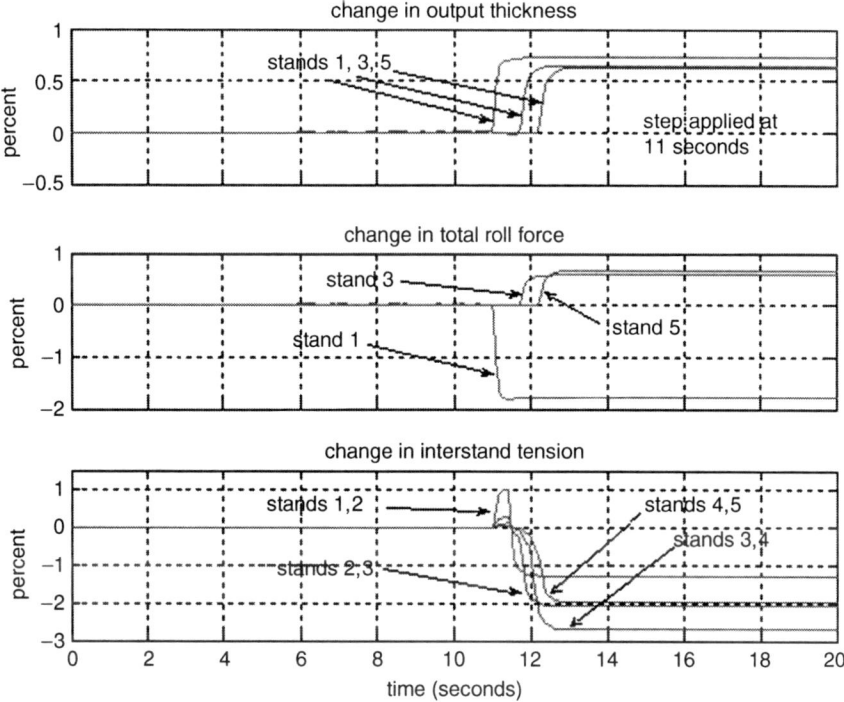

Fig. 2.5 Responses to +2% step change in stand 1 position actuator reference

conformed to [4], except where the changes in interstand tension were not large. These differences in smaller changes in interstand tensions are not considered significant. Step changes in the friction coefficients produced less significant changes in output thicknesses, total roll forces, and interstand tensions, and mostly are consistent with the results of the other two simulations.

The magnitudes of interstand tension showed greater discrepancies with Geddes than with Bryant in most cases (Table 2.5). This might be attributed to the differences between Bryant and Geddes in the estimates of the friction coefficients, as noted earlier.

In Figure 2.4 the percent change in the output thickness of stand 1 propagates through the downstream stands essentially without attenuation. Taking mass flow as constant, this is what would be expected intuitively assuming nearly constant forward slips, and conforms to the dynamic responses of both [4, 5]. The dynamic responses of the total roll forces (Figure 2.4) follow the changes in the stand output thicknesses. In Figure 2.4 the dynamic responses of the interstand tensions generally agree with those given in [4, 5].

Thus, considering the simulation results, the simplified model developed confirms reasonably well with the "benchmark" results of Bryant and also in general conforms to the results of Geddes and therefore is well-suited for the development

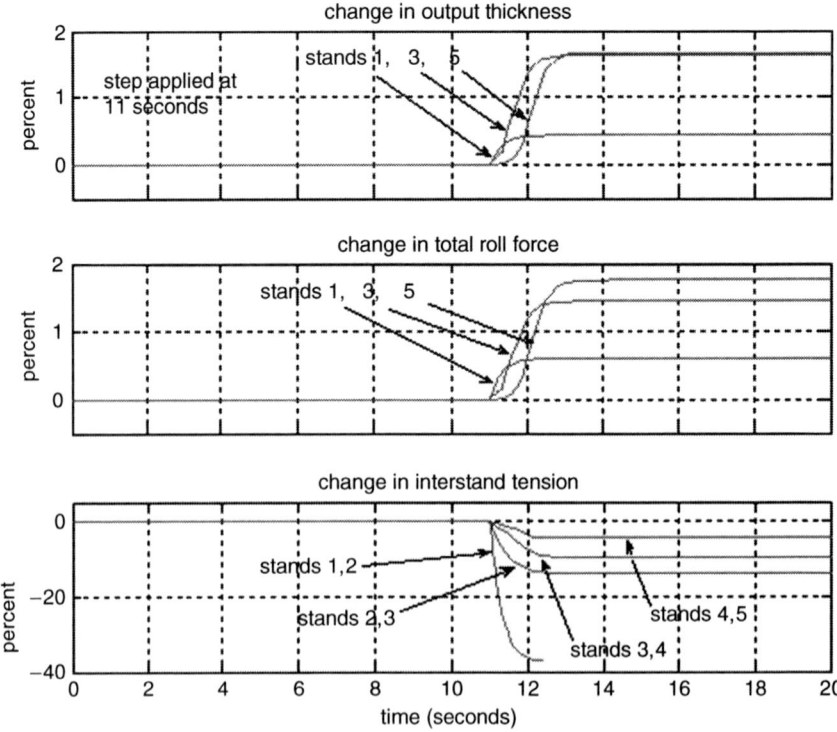

Fig. 2.6 Responses to +2% step change in the stand 1 speed actuator reference

and basic investigation of various techniques for the control of the tandem cold metal rolling process.

References

1. Roberts WL. Mathematical models related to rolling force. In: Cold rolling of steel. New York. Marcel Dekker; 1978.
2. Orowan E. The calculation of roll pressure in hot and cold flat rolling. Proc Inst Mech Eng. 1943;150(4):140–67.
3. Von Karman T. Beitrag zur theorie des walzvorganges. Z Angew Math Mech. 1925;5:139.
4. Bryant GF. The automation of tandem mills. London: The Iron and Steel Institute; 1973.
5. Geddes EJM. Tandem cold rolling and robust multivariable control. PhD thesis, Leicester: University of Leicester; 1998.
6. Guo RM. Analysis of dynamic behaviors of tandem cold mills using generalized dynamic and control equations. IEEE Trans Ind Appl. 2000;36(3):842–53.

Chapter 3
Conventional Control

3.1 Background

The tandem rolling of cold metal strip is a complex nonlinear multivariable process with stringent requirements on the finished product. The goal of a tandem cold metal rolling process is to produce flat sheets of metal of very high quality, where high quality implies conformance of the end product to a desired geometry, including strip flatness.[1] In an ideal cold rolling application no closed-loop control is required. This is because the settings of the various actuators of the mill are preset according to a pre-calculated rolling schedule which is based on the desired reduction at each stand, the desired interstand tensions, and the desired strip flatness. The rolling schedule calculations consider the loading capabilities of the various mill stands plus the capabilities and limitations of the mill actuators, which typically are the work roll position actuators, the work roll speed actuators, and the actuators (such as roll bending jacks) which affect the flatness of the strip. With ideal mill models and actuators, and in the absence of uncertainties and disturbances the rolling schedules can exactly predict the actuator settings so that open-loop control is sufficient to produce a high quality product.

However, as with almost all industrial processes the tandem cold rolling process is far from ideal, as there are numerous uncertainties and disturbances which make the application of closed-loop control essential to achieving the desired quality of the final output, and to assuring the stability of rolling by reducing excursions in interstand tensions and thicknesses. In addition to the disturbances and uncertainties, the process itself is highly complex and nonlinear, there are long time delays involved which change considerably with the mill speed, the mill characteristics can change during normal operations, a wide range of product must be accommodated with the product characteristics often changing significantly during the processing of a single product and between different products, plus the requirement that the control structure must present a user-friendly environment to the

---

[1]The terms "shape" and "flatness" sometimes are used interchangeably to imply the same condition of the strip. For purposes of this work, this understanding is adopted herein. More about this is considered in Section 3.6.

J. Pittner and M.A. Simaan, *Tandem Cold Metal Rolling Mill Control*,
Advances in Industrial Control, DOI 10.1007/978-0-85729-067-0_3,
© Springer-Verlag London Limited 2011

commissioning and operating personnel and promote physical intuition in the design process. As easily can be seen, all of these put together create an extremely challenging task for the control engineer.

To help meet this challenge, what is presented in this chapter is intended as background material to provide some insight into several basic concepts for the control of various area of the tandem cold rolling process. Certainly not every method of control can be addressed, and it is recognized that many of the control techniques presently in use go beyond the basics. However it is considered that what is furnished can provide a firm foundation for understanding the conventional methods and the more advanced concepts presented in Chapters 4 and 5, plus other advanced techniques developed for use in the various areas of tandem cold mill control.

3.2 Disturbances and Uncertainties

Major disturbances generally can be divided into two categories: (1) external disturbances, and (2) disturbances which are internal to the mill itself. Some major external disturbances arise mostly from the effects of previous processing in the hot rolling area where for example often a reheating furnace is employed to heat large metal slabs to a temperature suitable for further processing in a hot rolling mill. These disturbances are depicted in Figures 3.1 and 3.2, where the changes in the thickness and hardness of the incoming strip are caused by the hot slabs being in contact with their colder supports in the furnace, and with the higher frequency excursions being due to eccentricity-type effects in the hot mill rolls. These effects also depend on the speed of the mill as can be seen from the following figures for disturbances at typical values of thread speed and run speed.

The major internal disturbances are those generated as a result of eccentricity-type effects in the cold mill rolls. These effects are the result of axial deviations between the roll barrel and the roll neck caused by irregularities in the mill rolls, in the roll bearings, or in both, which produce cyclic variations in the strip thickness. The major contributors to roll eccentricity are the backup rolls in a mill stand, with the work rolls being a lesser contributor. More detail regarding roll eccentricity effects and the means for compensating for them is presented in Section 3.5 and later in Chapter 5.

The significant uncertainties also are of two types: (1) uncertainties in modeling and (2) uncertainties in measurement. These can result in undesirable deviations in the stand output thicknesses and in the interstand tensions from their values at an operating point, and thus an effective means of their mitigation is essential. In the model, there are usually significant uncertainties in: (1) the coefficient of friction of an individual stand, (2) the compressive yield stress (*i.e.* the hardness) of the strip, and (3) the modulus of elasticity (*i.e.* the mill modulus) of an individual mill stand.

Moreover, aside from these uncertainties there also are other deviations in the model from the physical plant (*i.e.*, the mill, the actuators, and the strip) which are

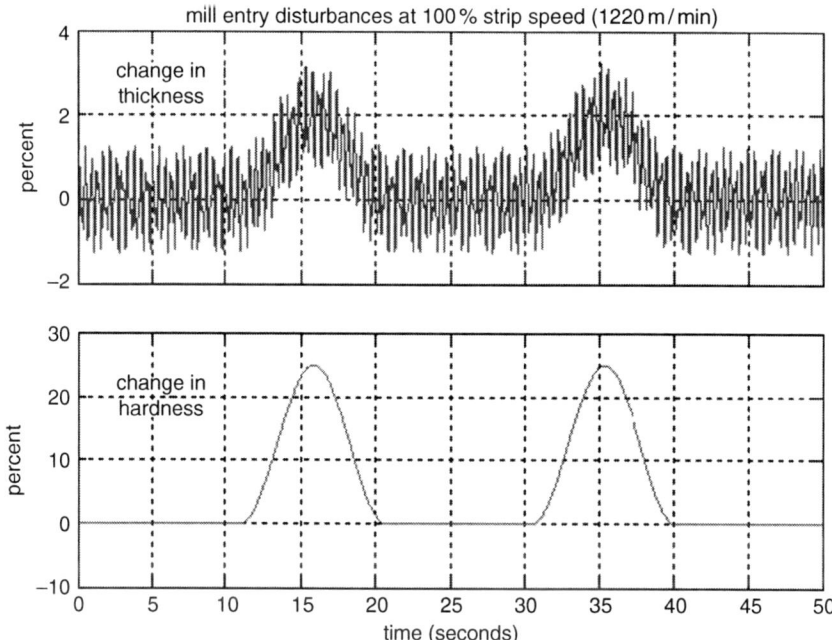

Fig. 3.1 External disturbances at typical run speed

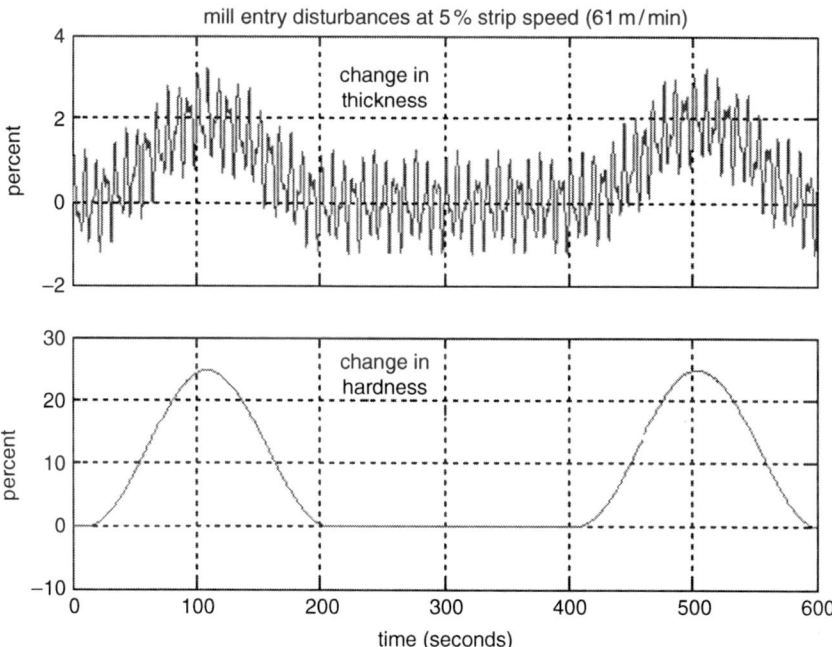

Fig. 3.2 External disturbances at typical thread speed

Table 3.1 Typical modeling
uncertainties

Parameter	Magnitude of percent uncertainty (%)
μ	20
$\bar{k}$	25
M	10

unavoidable since there is no model which can duplicate the plant exactly. These deviations however are less significant since most models (including the one described in Chapter 2) are verified against simulations which are generally accepted as "benchmarks" for model verification and against data from operating mills. Table 3.1 lists the individual uncertainties.

In Table 3.1, μ represents the friction coefficient, $\bar{k}$ is the mean compressive yield stress, and M is the mill modulus. The percentage listed for an individual entry represents the uncertainty in the initial estimate plus any uncertainty occurring during operation which could result from effects such as temperature change, roll wear, variations in lubrication parameters, or other conditions which are not explicitly described by the model.

The uncertainty given for μ is determined by best judgment using the data given in Roberts [1], considering that μ is derived empirically by evaluating the rolling lubricant in terms of two parameters designated as the first and second frictional characteristics, based on the effectiveness of the lubricant. This effectiveness depends on conditions such as the physical and chemical nature of the lubricant, the physical and chemical nature of the roll and the strip surfaces, the design of the mill stand, the mill speed, and the reduction. More detail related to estimation of μ and its uncertainty can be found in the works of Roberts and the references cited therein, and also in Lenard [2].

The uncertainty denoted for average strip compressive yield stress $\bar{k}$ is based on empirical data for mild steel produced on operating mills, plus estimates based on experience. In some instances however the estimate of uncertainty in $\bar{k}$ is even greater than what is shown in Table 3.1, and can depend on the material being processed.

While the mill modulus M is inferred to be constant in many references, in fact it depends heavily on the backup roll diameter and varies as the backup roll diameter changes with temperature and roll wear, and also with other effects such as the strip width. As will be shown later in this chapter, this uncertainty is significant in that changes in the mill modulus can cause corresponding changes in the output thickness which can seriously affect the performance of the controller, and therefore should not be overlooked in the development of a control strategy. A more thorough treatment of some of the causes of mill modulus uncertainties can be found in [3, 4].

The uncertainties in the measurement of process variables important for control are listed in Table 3.2. The uncertainties listed are typical as derived from recent manufacturer's data and includes all sources of error in the measurement available at the controller. The uncertainty for the thickness measurement is listed as zero based on the assumption of a suitable calibration so that any offset caused by

Table 3.2 Typical measurement uncertainties

Parameter	Magnitude of percent uncertainty (%)
$h_{out1m(5m)}$	0
F	0.2
T	0.2
S	0.05
V	0.1
$V_{in,out,i}$	0.025

error is eliminated and only deviations around the operating point are given. This is generally consistent with data presented from operating mills for this measurement.

The listings for uncertainties in other measurements in Table 3.2 are percentages of measured values except for total roll force F, interstand tension force T, and position actuator position S which are percentages of full scale values. The listing for V includes the uncertainty in the work roll diameter. The listing $V_{in,out,i}$ for strip speeds at the input (or output) at a mill stand is based on the use of reliable high accuracy laser-based velocimeters. These measurements of strip speed can be used as inputs to the controller. More about the use of these measurements to improve controller performance is presented in Chapter 5.

3.3 Other Considerations

In addition to maintaining good performance in the presence of disturbances and uncertainties, which includes keeping tight tolerance in the centerline thickness at the mill output, the mill controller must be capable of performing other functions and handling other conditions encountered during normal operation. These include: (1) maintaining good performance during mill acceleration and deceleration, (2) reducing excursions in interstand thickness and tension, (3) providing for the operator to adjust independently the strip thickness and tension anywhere in the mill, (4) accommodating for continuous mills a very rapid change in product, (5) control of the strip flatness, (6) control during threading, and (7) provision of a control structure that is easy to implement for design and commissioning personnel who have a limited background in advanced control theory. In addition, the controller algorithms must be capable of running on hardware and software platforms sufficiently fast to properly control the process.

For a stand-alone mill acceleration from thread speed to run speed and the reverse is a part of normal operation during coil changes. In a continuous mill, often the process is slowed from run speed to a speed suitable for weld transfer to reduce the likelihood of strip breakage during passage of the weld, and to be at a speed compatible for the cutting of the strip by the shear with a subsequent transfer of the oncoming strip to the next available coiler at the mill exit.

Maintaining low excursions in interstand thicknesses and tensions over all regimes of operation is important to support the stability of rolling so that the likelihood of cobbles (*i.e.* wrecks in the mill) is reduced. Additionally, during normal operations the mill operator often makes slight adjustments in the reduction taken at a stand to slightly shift the loading at each of the stands in the mill, and also may adjust individually the interstand tensions.

In the case of continuous mills, the product can change very rapidly (in milliseconds) during the passage of the weld through the roll bite. This puts severe requirements on the mill controller to reduce the rapid change in the roll force during weld passage so as to prevent damage to the work rolls. Also, the controller must control the change in referencing from the present strip to the next strip in a manner that reduces the length of off-gage material near the weld during the transfer, assure that during the transfer the mass flow of the material is maintained to support the stability of rolling, and preclude severe excursions in the strip tension. Additional material related to the control of continuous mills during the weld passage is provided in Section 3.4.6.

Control of strip flatness will be briefly addressed in Section 3.6, and control during threading will be briefly addressed in Section 3.7. However, unless noted otherwise, the material presented herein will assume that the mill is successfully threaded which also implies that tension is established throughout the mill.

3.4   Basic Conventional Concepts

In this section, some basic conventional concepts are presented for control of the following variables in the tandem cold mill: the interstand tensions, the strip thicknesses at the outputs of the individual mill stands, and in certain cases the total roll force. The devices that are actuated by the controller to effect the control efforts are assumed to be hydraulic cylinders for the setting the roll gap positions, and speed controlled electric drives for setting the work roll speeds. It is assumed that the controller for strip thicknesses and tensions is essentially non-interacting with the controller for the strip flatness. This is based on experience and material presented in Roberts [5] which gives criteria for considering these two controllers as non-interacting, *i.e.* each controller can be treated separately. What is given in Roberts is derived from a study and experimentation performed by Sabatini and Tarokh (see [5]), and is generally consistent with current practice, although in certain instances the two controllers are combined into a single configuration.

Each of the mill stands with its associated interstand tension area is a multivariable system, with coupling to other stands through the interstand tension and the exit thickness via a time delay. The coupled stands then become an overall larger multivariable system with interacting variables between the individual areas, so that the perturbation of a variable at one location in the mill will affect other variables throughout the mill. The magnitude of the interactive effects will vary depending on the strengths of the coupling between the various variables, as some couplings are much stronger than others.

As an example of the interactions between mill variables, a change in the interstand tension at a specific location will be seen after an almost negligible time delay in other tensions, as the tension perturbation travels in the metal strip as the speed of sound. In the case of a perturbation in the strip thickness, a change will be seen almost instantly in the interstand tensions, but with the effects of the interstand time delays as the change in thickness moves through the downstream stands. This can be seen in Figure 2.4 where a step change in the input thickness causes an almost instantaneous change in interstand tensions throughout the mill, but the responses in thicknesses are delayed essentially by the interstand time delays. In most cases, these types of interactions between mill variables are exploited in determining the structure of industrial controllers. Thus the basic approach to conventional industrial control of this system has been to use a combination of single-input-single-output (SISO) control loops and single-input-multi-output (SIMO) control loops, with the control strategy determined based on the consideration of these types of interactions. Generally the conventional control of interstand tension and strip thickness is based on this type of approach.

3.4.1 Non-Interactive Control

The concept of non-interactive control was developed by Bryant and others [6] as a means to assure that any thickness or tension in the mill can be adjusted independently without significantly affecting the other thicknesses or tensions. This concept, while not applied to its fullest extent in most conventional controllers and relies on several unrealistic assumptions which are addressed in what follows, nonetheless forms the underlying basis for many of the conventional control approaches which are described in the literature or implemented in actual practice and therefore is deemed worthy of brief consideration in this chapter.

The non-interactive control requires a process model that is linearized around an operating point established by the rolling schedule. The operating point is assumed to be an equilibrium point. Of interest in this linearization are the changes in specific roll force P and the forward slip f. Specifically,

$$\delta P_i = \delta h_{in,i}\varepsilon_1 + \delta h_{out,i}\varepsilon_2 + \delta T_{i-1}, i\varepsilon_3 + \delta T_{i,i+1}\varepsilon_4, \tag{3.1}$$

$$\delta f_i = \delta h_{in,i}\varepsilon_5 + \delta h_{out,i}\varepsilon_6 + \delta T_{i-1,i}\varepsilon_7 + \delta T_{i,i+1}\varepsilon_8, \tag{3.2}$$

$$\delta P_{i+1} = \delta h_{in,i+1}\varepsilon_9 + \delta h_{out,i+1}\varepsilon_{10} + \delta T_{i,i+1}\varepsilon_{11} + \delta T_{i+1,i+2}\varepsilon_{12}, \tag{3.3}$$

$$\delta f_{i+1} = \delta h_{in,i+1}\varepsilon_{13} + \delta h_{out,i+1}\varepsilon_{14} + \delta T_{i,i+1}\varepsilon_{15} + \delta T_{i+1,i+2}\varepsilon_{16}, \tag{3.4}$$

where h_{in} is the input thickness, h_{out} is the output thickness, T is the strip tension force, and i is the stand number. The coefficients ε_j ($j = 1$–16) are defined as the partial derivatives evaluated at the operating point, and which are associated with the applicable variables,

$$\varepsilon_1 = \frac{\partial P_i}{\partial h_{in,i}}, \varepsilon_2 = \frac{\partial P_i}{\partial h_{out,i}}, \varepsilon_3 = \frac{\partial P_i}{\partial T_{i-1,i}}, \varepsilon_4 = \frac{\partial P_i}{\partial T_{i,i+1}}, \tag{3.5}$$

$$\varepsilon_5 = \frac{\partial f_i}{\partial h_{in,i}}, \varepsilon_6 = \frac{\partial f_i}{\partial h_{out,i}}, \varepsilon_7 = \frac{\partial f_i}{\partial T_{i-1,i}}, \varepsilon_8 = \frac{\partial f_i}{\partial T_{i,i+1}}, \tag{3.6}$$

$$\varepsilon_9 = \frac{\partial P_{i+1}}{\partial h_{in,i+1}}, \varepsilon_{10} = \frac{\partial P_{i+1}}{\partial h_{out,i+1}}, \varepsilon_{11} = \frac{\partial P_{i+1}}{\partial T_{i,i+1}}, \varepsilon_{12} = \frac{\partial P_{i+1}}{\partial T_{i+1,i+2}}, \tag{3.7}$$

$$\varepsilon_{13} = \frac{\partial f_{i+1}}{\partial h_{in,i+1}}, \varepsilon_{14} = \frac{\partial f_{i+1}}{\partial h_{out,i+1}}, \varepsilon_{15} = \frac{\partial f_{i+1}}{\partial T_{i,i+1}}, \varepsilon_{16} = \frac{\partial f_{i+1}}{\partial T_{i+1,i+2}}. \tag{3.8}$$

Using (2.19) and (2.16) for the forward slip and output thickness gives

$$\delta V_{out,i} = \delta V_i(1 + f_i) + V_i \delta f_i, \text{ and} \tag{3.9}$$

$$\delta S_i = \delta h_{out,i} - \frac{W \delta P_i}{M}, \tag{3.10}$$

where symbols are as previously defined in Chapter 2. Assuming no change in the strip width W across the roll gap, the steady-state mass flow relationship results in

$$V_{in,i} h_{in,i} = V_{out,i} h_{out,i}. \tag{3.11}$$

Where there is a change in mass flow with constant strip width, using (3.11) the change can be represented as

$$\delta V_{in,i} h_{in,i} + V_{in,i} \delta h_{in,i} = \delta V_{out,i} h_{out,i} + V_{out,i} \delta h_{out,i}. \tag{3.12}$$

As a simple example in the application of (3.12), with $\delta h_{in,i} = 0$ and $\delta V_{out,i} = 0$, and applying (3.11) gives

$$\delta V_{in,i} = V_{in,i} \frac{\delta h_{out,i}}{h_{out,i}}. \tag{3.13}$$

Also in steady-state conditions,

$$h_{out,i-1} = h_{in,i}. \tag{3.14}$$

3.4.1.1   Control of Interstand Tension

Using the above derived relationships, it can easily be seen that under certain assumptions for a change in tension $\delta T_{i,i+1}$, controlling the roll gap actuator positions and work roll speeds on stands i and $i + 1$ can tend to prevent excursions in the other thicknesses and tensions. This is equivalent to computing the changes in actuator references to achieve only the change in tension $\delta T_{i,i+1}$. For this case, the following are applicable under steady-state conditions:

$$\delta h_{in,i} = \delta h_{out,i} = 0, \tag{3.15}$$

$$\delta V_{in,i} = \delta V_{out,i} = 0, \text{ for all } i, \text{ and} \tag{3.16}$$

$$\delta T_{j-1,j} = 0, \text{ for all } j \neq i + 1. \tag{3.17}$$

Applying (3.9), (3.10), (3.15)–(3.17) into (3.1)–(3.4) results in

$$\delta S_i = -\varepsilon_4 \frac{W \delta T_{i,i+1}}{M}, \tag{3.18}$$

$$\delta S_{i+1} = -\varepsilon_{11} \frac{W \delta T_{i,i+1}}{M}, \tag{3.19}$$

$$\delta V_i = -\varepsilon_8 \frac{V_i}{(1 + f_i)} \delta T_{i,i+1}, \tag{3.20}$$

$$\delta V_{i+1} = -\varepsilon_{15} \frac{V_{i+1}}{(1 + f_{i+1})} \delta T_{i,i+1}, \tag{3.21}$$

which are the references for the gap actuators and work roll speed actuators for stands i and $i + 1$, with the references for all other actuators remaining unchanged.

3.4.1.2   Control of Interstand Strip Thickness

A similar but a slightly more complex non-interactive approach can be applied to develop a method for the control of interstand strip thickness. The main idea is to determine the actuator references needed to change the thickness at a stand output without changing the interstand tensions, and to keep the change in the thickness from propagating throughout the mill.

An insight into this approach can be obtained by considering that (1) mass flow must be conserved throughout the mill, and (2) the input thickness is to remain unchanged throughout the mill, (*i.e.*, $\delta h_{in} = 0$). At stand i by conservation of mass flow, and with $\delta V_{out,i} = 0$,

$$\delta V_{in,i} = V_{in,i} \frac{\delta h_{out,i}}{h_{out,i}}. \tag{3.22}$$

At the mill stands upstream of stand i, since mass flow must be conserved and the input thickness must remain unchanged,

$$\delta V_{in,j} = V_{in,j} \frac{\delta h_{out,i}}{h_{out,i}}, \tag{3.23}$$

where $(j < i)$ denotes the upstream stands.

However, the output thickness at upstream stand j also must remain unchanged which gives

$$\delta V_{out,j} = V_{out,j} \frac{\delta h_{out,i}}{h_{out,i}}. \tag{3.24}$$

For the stands downstream of stand i at the time of the change in the output thickness at stand i, mass flow at these stands remains unchanged, which implies that the output thicknesses at the downstream stands remain unchanged, and the input and output strip speeds remain unchanged, and therefore there is no adjustment required in the roll gap position actuators or in the work roll speed actuators.

At stand i the change in the output thickness causes changes in the specific roll force and the forward slip,

$$\delta P_i = \delta h_{out,i} \varepsilon_2, \tag{3.25}$$

$$\delta f_i = \delta h_{out,i} \varepsilon_6. \tag{3.26}$$

Compensation for these changes at stand i is made by adjusting the roll gap position and work roll speed actuators,

$$\delta S_i = \delta h_{out,i} \left(1 - \frac{W \varepsilon_2}{M} \right), \tag{3.27}$$

and

$$\delta V_i = -\delta h_{out,i} \varepsilon_6 \left(\frac{V_i}{1 + f_i} \right). \tag{3.28}$$

At the stands upstream of stand i the conservation of mass flow requires no adjustment to the roll gap position actuators, with an adjustment to the work roll speed actuators at the upstream stands as

$$\delta V_j = V_j \frac{\delta h_{out,i}}{h_{out,i}}, \tag{3.29}$$

where $(j < i)$ denotes the upstream stands.

After a time delay τ the change in the output thickness of stand i is seen at stand $i + 1$ as a change in the input thickness, *i.e.*

$$\delta h_{in,i+1} = \delta h_{out,i} e^{-\tau s}, \tag{3.30}$$

where s is the Laplace operator. In a manner similar to that for stand i, compensation is provided in the roll gap position and work roll speed actuators at stand $i + 1$ as

$$\delta S_{i+1} = \delta h_{out,i} e^{-\tau s} \varepsilon_9 \frac{W}{M}, \tag{3.31}$$

$$\delta V_{i+1} = -\delta h_{out,i} e^{-\tau s} \varepsilon_{13} \left(\frac{V_{i+1}}{1 + f_{i+1}} \right), \tag{3.32}$$

where ε_9 and ε_{13} are the coefficients for stand $i + 1$ corresponding to $h_{in,i+1}$. At stands downstream from stand $i + 1$ a change to the work roll speed is not required. At stands upstream of stand $i + 1$ the necessary change to the work roll speed is

$$\delta V_j = -V_j \frac{\delta h_{out,i} e^{-\tau s}}{h_{in,i+1}}, \tag{3.33}$$

where $(j < i + 1)$.

Examination of the previous structure of the non-interactive controller illustrates the complex interactions between the various variables in the tandem cold mill and provides good insight into their behavior in the presence of a tension disturbance and a thickness disturbance, plus illustrating the need for consideration of the conservation of mass flow in the various areas of the mill. However, while an examination of the non-interacting structure is a good exercise to become initially acquainted with the mill behavior, it should be recognized that several assumptions are made in the preceding analysis that would make this technique less than desirable in a practical setting.

A major assumption is that of assuming a model that accurately represents the process, and thus allows the usage of an open-loop control strategy. As noted previously in Section 3.1, this is highly unrealistic because of the many uncertainties that change under the various conditions occurring during mill operation. Moreover, the analysis is based on a linearized model which adds complexity to the controller design, especially considering the potential need for some adjustment in the linearizing coefficients while rolling to assure good control performance over the entire range of mill operation, and with significant variations in the product being processed. Additionally, it is assumed unrealistically that there is ideal matching between similar actuators throughout the mill, that an ideal determination of variables such as thicknesses and forward slips is available to the controller, and that a system for ideally tracking changes in thickness is in place and operational. However, as noted previously many of the conventional control strategies use some of the concepts developed in the non-interactive controller as can be seen in several of the control strategies described in the literature and in those that are in actual usage. This can be noted in the following presentation of certain approaches which are basic to conventional control.

3.4.2 Interstand Tension Control

As can be seen from the open-loop steady-state responses given for example in Tables 2.13 and 2.14 of Chapter 2, a small step change in the position actuator reference of a particular mill stand has a significant effect on the stand entry tension force, but a lesser effect on the exit tension of the stand. A typical basic conventional concept for the control of interstand tension force is denoted as "tension by gap" and is depicted schematically in Figure 3.3.

The dynamic open-loop response in interstand tension to a +5% step change in the reference of the position controller for the stand 3 roll gap position actuator at 100% speed is shown in Figure 3.4. The response is sufficiently fast to be suitable for tension control, so that this characteristic lends itself to controlling interstand tension by control of the position of the downstream roll gap. Responses at lower mill speeds are nearly identical. In these cases it is assumed that the stands involved are at or above a predetermined low speed. From zero speed up to this speed, generally the interstand tension is controlled by adjustment of work roll speed, with

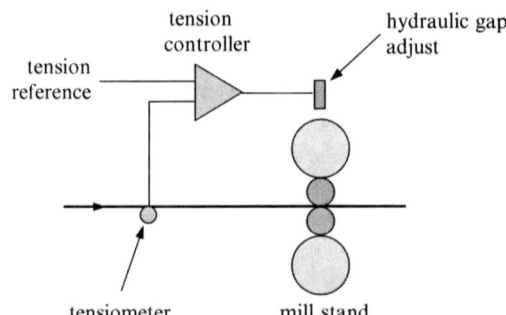

Fig. 3.3 Schematic for basic conventional control of tension by gap

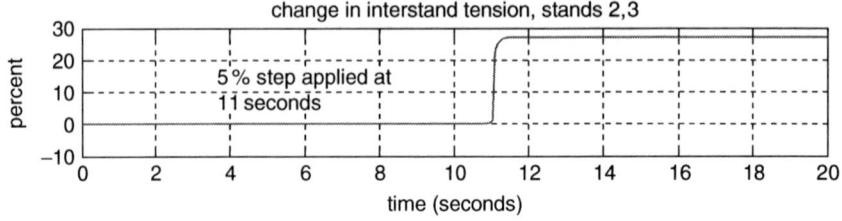

Fig. 3.4 Open-loop response of interstand tension to a +5% step change in the reference of the postion controller of the stand 3 roll gap position actuator at 100% speed

a cross-over at the predetermined speed to controlling the tension by adjustment of the roll gap. This approach, denoted as "tension by speed/tension by gap" is a basic method of conventional control. This dual approach to tension control is necessary since the mill must be at or above a predetermined speed greater than zero to effect an acceptable change in the tension by a change in the roll gap setting.

Assuming a controller based on non-interactive techniques, some additional characteristics of the tension by gap method that are beneficial are that (1) the correction to an excursion in tension caused by a deviation in thickness inherently tends to correct the thickness deviation, essentially irrespective of the source of the deviation, and (2) the tension by gap approach is user-friendly and quite easy to tune at commissioning.

Another method of basic conventional control of interstand tension force is "tension by speed". This approach is sometimes used in applications wherein the roll positioning control is used for control of thickness, or for other purposes. In this method, the difference between the strip speed at the input to stand $i + 1$ and the strip speed at the output of stand i is adjusted by changing the speed of the work rolls at one of the stands. The applicable relationship for the interstand tension force is approximated using (2.25) as

$$\frac{d(T_{i,i+1})}{dt} \equiv \dot{T}_{i,i+1} = k(V_{in,i+1} - V_{out,i}), T_{i,i+1}(0) = T_{0,i,i+1}, \qquad (3.34)$$

where T is the interstand tension force between stands i and $i + 1$, where k is taken for this approximation as a constant that depends on Young's modulus, the distance between the centerlines of the two-stands, and the width and thickness of the strip in the interstand area, with other symbols as noted in (2.25). A schematic for the control of "tension by speed" is shown in Figure 3.5.

The dynamic open-loop response in interstand tension to a $+1\%$ step change in the reference of the stand 2 work roll drive at 100% speed is shown in Figure 3.6, with the work roll speed of stand 3 remaining fixed at 100%. The response depicting the difference in the strip speed at the input to stand 3 and the strip speed at the output of stand 2 is also presented to show in accordance with (3.34)

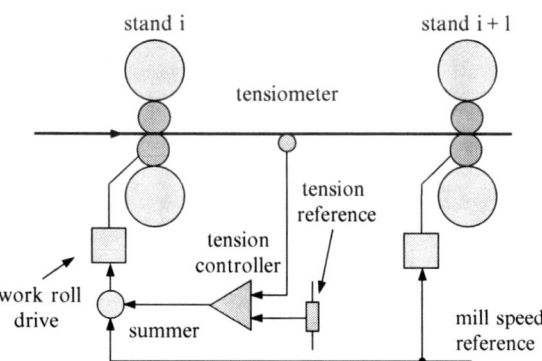

Fig. 3.5 Schematic for basic conventional control of tension by speed

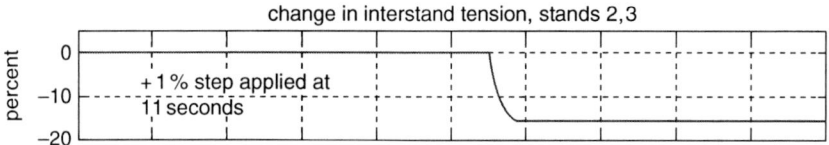

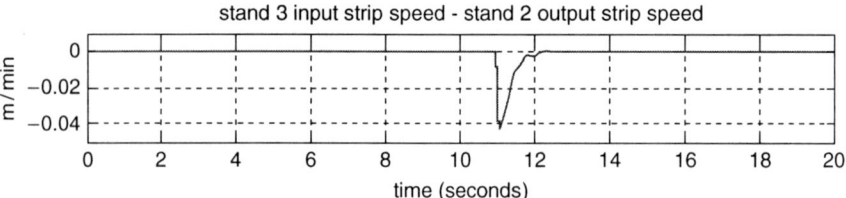

Fig. 3.6 Open-loop response of interstand tension and strip speed difference to a +1% step change in the reference of the stand 2 work roll speed actuator at 100% speed

how at steady-state these speeds must match, assuming that there is no stretching of the strip. As in the case of tension by gap, it can be seen that the open-loop tension response lends itself well for the control of tension by speed. In addition, the tension by speed method is user-friendly and easy to tune at commissioning.

3.4.3 Strip Thickness Control

There are many upon many techniques for automatically controlling the thickness (*i.e.* the gage) at the centerline of the strip as it exits the mill, and also the thicknesses at the exit of the intermediate stands. Of course, not every conceivable configuration can be evaluated. Rather than try to review a host of different specific configurations and then determine their strengths and weaknesses, it is preferred to present some basic concepts by which almost any particular approach to automatic thickness control (or automatic gage control, denoted as AGC) can be examined, and then present some applicable examples, so that the reader may get the general flavor of conventional automatic thickness control, and then use the previous material on non-interactive control plus what is presented herein to understand and evaluate a particular application.

The conventional control of strip thickness generally involves a few basic concepts that usually are combined in some fashion to implement an overall control approach. In addition to the non-interactive control concepts as presented previously these are: (1) the use of feedback, (2) the use of feedforward, (3) the control of mass flow, and (4) the use of BISRA measurements to infer strip thickness.

Unless otherwise noted, in what follows it will be assumed that the basic conventional control strategy will be initially for a stand-alone five-stand mill with a thickness gage at the exit of the first and last stands. Other configurations will be addressed in the examples which follow in later subsections.

3.4.3.1 Use of Feedback

Feedback control generally implies direct measurement of the strip centerline thickness by a suitable gaging method, with the measurement usually being made by X-ray based technology, or occasionally by isotope gages for thickness measurement of heavier strips. As noted previously, these thickness measurements generally are located at the exit of the first and last stand of a stand-alone mill, with a third additional measurement at the entry of the first stand in the case of a continuous mill. The number of these measurements is limited because reliable thickness measurement systems usually are expensive and require additional physical space that often is unavailable between the adjacent stands of the mill, especially in the case of an upgrade of an existing installation where the interstand dimensions are often fixed by the existing locations of the stands. However, this is not always the case, and in several instances thickness measurements are located at certain intermediate stands, where provision is made for their installation at the time of the physical layout of the mill, or as part of the upgrade where feasible. In every application the thickness gage locations must be carefully and thoroughly considered during the design phase, and confirmed by multiple simulations to address their usefulness in improving the controller performance.

A difficulty with the thickness measurement is that there is an inherent time delay between the strip thickness as seen at the exit of a stand and the strip thickness as measured by the thickness gage because of the physical distance (about 1–2 m) between the stand and the gage. This can be significant at lower speeds as the delay becomes much longer than at the higher speeds; for example the thread speed may be about 5% of the run speed which lengthens the delay considerably.

The thickness measurement at the exit of the first stand provides feedback to a control loop that uses the roll gap actuator to adjust the thickness out of this stand. This provides a reasonably fast correction to thickness errors at the entry end of the mill. At the exit end of the mill, in a basic conventional method the thickness error is determined by using the thickness measurement. Assuming that the strip tension is controlled by the "tension by gap" method, the thickness error then is fed back to be applied as a trim on the speeds of some of the upstream stands. An exception is one of the stands (often the second, third, or fourth stand in a five-stand mill) which is denoted as the "pivot stand" whose speed is not trimmed, and acts as the mill pace setter for the various regimes of operation such as speed change from thread to run and the reverse. However, the thickness control loop by itself is slow in responding to excursions from the desired thickness at the mill output as it must change the speeds of the work rolls over a long length of the strip before a correction can be fully achieved. Its usefulness therefore is mostly in removing any long-term error in the incoming strip from the hot mill.

3.4.3.2 Use of Feedforward

The thickness measurement at the exit of the first stand is tracked through the mill from the thickness gage to certain of the downstream stands and applied as a feedforward trim on the work roll speeds, in conjunction with the feedback signals noted above from the thickness measurement at the mill exit. In this way incoming excursions in the entry thickness can be mitigated to some extent. However, a major weakness in this approach is that changes in the hardness of the strip are not seen in the downstream stands, and these changes can have an effect on the associated thicknesses.

3.4.3.3 Control of Mass Flow

The control of mass flow is a very important concept in the control of tandem cold rolling. Usually the mass flow concept is applied assuming that the strip width is constant throughout the mill, so that in steady-state, since the mass flow is invariant over the entire mill, it can be expressed anywhere in the mill as

$$MF = vh, \tag{3.35}$$

where v is the strip speed and h is the thickness at a particular point of interest, which is usually at the input or output of a mill stand. A typical implementation is to attempt to hold the strip thickness constant by holding the mass flow constant at a particular stand and then maintaining the strip speed constant at another stand, which then results in almost constant thickness. As noted previously, in a typical conventionally controlled mill, one of the stands is chosen as a "pivot stand" whose speed remains untrimmed and is used as a pace setter for the remainder of the mill. Using the thickness measurement at the mill exit, the speeds of certain of the mill stands can be adjusted, and based on the constant speed of the pivot stand with conservation of mass flow, an exit thicknesses is produced that is approximately constant. However, as also noted previously this approach has the drawback of a long time delay to correct for excursions in thickness and therefore has limited usefulness to correct for faster excursions. This is especially significant during speed changes where faster corrections are needed to reduce undesireable excursions in thickness.

Another and more effective use of the mass flow approach is to measure the strip thickness near the output of the first stand, and then track the measured thickness to the input of the second stand using a reliable high accuracy speed sensor for more accurate tracking and speed measurement. Since both the strip speed and the strip thickness are known at the input of the second stand, a measurement of the strip speed at the output of the second stand will give a good estimate of the strip thickness at the stand output, by using the conservation of mass flow across the roll bite to compute the output thickness,

$$h_{out,2} = h_{in,2} \frac{V_{in,2}}{V_{out,2}} k_{cf}, \tag{3.36}$$

where k_{cf} is a factor for minor corrections depending on conditions such as small changes in strip width. The estimated strip thickness then can be tracked to the next stand and the estimate repeated, and subsequently throughout the mill. While (3.36) represents a steady-state relationship across the roll bite, the error in ignoring any dynamics involved is very low.

It should be noted that this method is implemented with current technology such as reliable high-accuracy laser speed sensors. However, these sensors require additional maintenance which can be somewhat of a drawback which usually is justified based on the improved performance in control of the interstand thicknesses.

It also is quite important to maintain constant mass flow to assure the stability of rolling. This is of special significance in a continuous application wherein the strip characteristics will change in milliseconds as the strip of an oncoming coil goes through the roll bite of a mill stand. More about the methods for maintaining constant mass flow in various areas of the continuous mill under these conditions will be presented later in this chapter and in Chapter 5.

3.4.3.4   Use of BISRA (Gagemeter) Measurements

The BISRA relationship is a mathematical relationship that has been developed by the British Iron and Steel Research Association (BISRA) in the early 1950s. This rela tionship (more often referred to as "gaugemeter" or "BISRA gaugemeter") has been used to estimate the thickness of the strip exiting the roll bite without requiring a direct measurement. Reliable direct measurement of strip thickness at the exit of the roll bite requires expensive and complex equipment, and therefore in general is not usually done, except at the exit of the first and last stands in a stand-alone mill, with an additional measurement at the mill entry for a continuous mill. However, for intermediate stands an estimate of the output thickness can be obtained by measuring the rolling load and using the mill stretch characteristic (Figure 3.7).

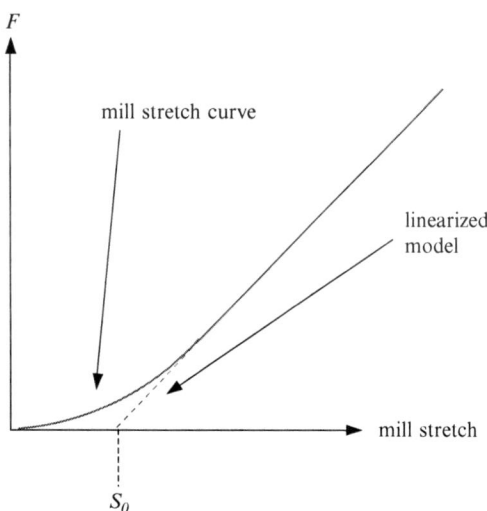

Fig. 3.7 Mill stretch characteristic

The mill stretch characteristic is a relationship that treats the mill as a spring which stretches according to the rolling load. Since most of the rolling operation occurs where this characteristic is nearly linear, the following linearized model of this relationship often is used as a suitable approximation,

$$mill\ stretch = S_0 + \frac{F}{M},\tag{3.37}$$

where S_o is the extrapolated intercept of the linearized model on the mill stretch axis, F is the total roll force, and M is the mill modulus. If the mill rolls are initially separated by the dimension of the unloaded roll gap, then under actual operating conditions (*i.e.* strip in the roll bite), the stand output thickness h_{out} as depicted in Figure 3.8 is given by

$$h_{out} = S + S_0 + \frac{F}{M}.\tag{3.38}$$

The mill stretch curve is based on a calibration that usually is performed after each roll change. As can be seen from Figure 3.8, there is a nearly linear region approximated by (3.38) and a nonlinear region, which is sometimes approximated by a polynomial function, or handled in the controller by other appropriate methods. It should be noted that the estimation of the output thickness h_{out} is sensitive to uncertainties in the mill modulus M. These uncertainties can arise due to changes in the diameter of the backup rolls that are caused by heating and mechanical wear, plus other effects. For example, in the nearly linear region in a typical application an uncertainty of 10% in M results in an error in the estimation of h_{out} of about 9% for a nominal strip output thickness of 1.6 mm, which can result in a corresponding error in a controller which uses this estimation to represent the actual strip thickness.

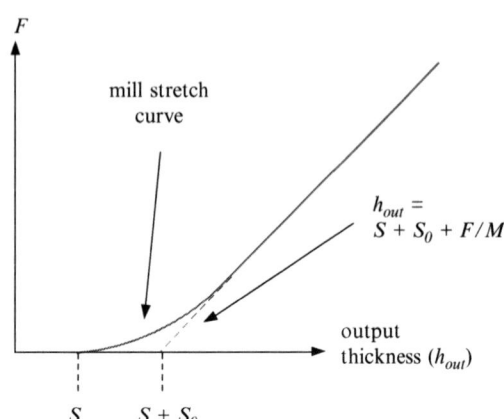

Fig. 3.8 Mill stand output characteristic

Figure 3.9 depicts the basic concept for use of the gagemeter method for control of the stand output thickness. In Figure 3.9 the mill stand characteristics are shown with the characteristics of the strip, so that the operating point denoted as A_1 results in a total roll force of F_1 and produces and output thickness of $h_{out,1}$, with an unloaded roll gap of S_a. Should the strip characteristic change (*e.g.* a change in hardness or in incoming thickness) so that the operating point is shifted from A_1 to A_2, the total roll force will shift from F_1 to F_2 and the output thickness will change from $h_{out,1}$ to $h_{out,2}$. To compensate for this change in h_{out}, the work roll gap actuator can be moved equivalently to an unloaded roll gap movement of S_a to S_b to regain the desired output thickness $h_{out,1}$, with a corresponding change in total roll force from F_2 to F_3, and a new operating point A_3. Thus the motion of the hydraulic cylinder controlling the roll gap position is such that its final position fully compensates for the mill stretch and produces the original output thickness $h_{out,1}$. In this particular case, *i.e.* where there is full compensation, the gagemeter method is said to produce what is termed an "infinitely stiff" mill.

However, due to eccentricity-like effects in the mill rolls there can be an increase in the rolling load with a corresponding imprinting in the thickness of the strip, so that these eccentricity-like effects can get rolled into the strip thickness. Depending on the strip hardness, the magnitude of these imprints can be significant. Further, where the BISRA relationship is used to develop a thickness feedback signal for closed-loop control, the eccentricity-like effects will be interpreted as an increase in the output thickness, when in fact the output thickness actually decreases, so that the thickness controller acts to move the roll gap position actuator in the wrong direction, with the disadvantage that the eccentricity effect is worsened rather than mitigated.

To attempt to reduce some of the imprinting, full compensation is often not implemented, and there is a compromise between fully mitigating the effects of changes in output thickness due to changes in input thickness and reducing the imprinting of the strip due to eccentricity. In addition, to overcome the effects of eccentricity in the mill rolls and the disadvantages of uncertainties in M, various means to mitigate these effects have been developed and implemented. Some of these methods are addressed further in Chapter 5.

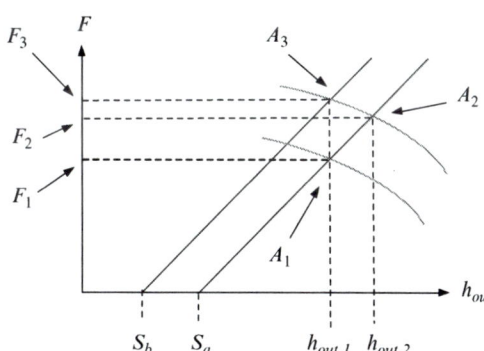

Fig. 3.9 Compensation for changes in the strip characteristic

3.4.4   An Illustrative Tandem Cold Mill Control Structure

Using the above described conventional methods, a basic control structure can be developed that illustrates the usage of these methods. The control structure is similar to one described in [7], and is close to what is used in certain actual applications as described for example in Section 3.4.5. The intent of presenting this structure is to give some idea at a very basic level of the manner in which many of the conventional approaches described previously can be implemented. Using this illustrative structure as a guide, it is expected that the reader can extrapolate some of the concepts presented to understand and evaluate other conventional and more advanced applications, some of which are presented as examples of typical installations later in this chapter and in Chapters 4 and 5. The illustrative structure is presented in Figure 3.10.

As can be seen in Figure 3.10, the interstand tensions upstream of stands 2 through 4 are controlled as tension by gap. At stand 5, the tension is controlled by trimming the work roll speed of this stand. The function of stand 5 in this configuration is to control the roll force and thus the surface finish of the end product.

The control of thickness is by feedforward and feedback loops using thickness measurements at the mill exit and just after stand 1. The feedback loop is a single-input-multi-output configuration as previously described wherein an output thickness trim on the work roll speeds of certain mill stands is computed to correct for thickness errors based on the measured thickness at the mill output. This is a slow-responding loop and is intended for long term corrections to the exit thickness, wherein the thickness errors of this long-term type originate in an upstream hot rolling process. The idea of this approach is to use mass flow techniques, so that an excursion in exit thickness from the desired value is seen as a change in the mass flow, since the speed of stand 4 is fixed. This excursion is fed back to stands 1 through 3 as trims on their speed references. However, the speed of stand 4 is fixed as it is the pivot stand, and thus the thickness at the output of stand 4 must change to conserve the mass flow, and thus offer some correction to the error in thickness. It should be noted that the reduction taken by stand 5 is very low and therefore has little effect on the output thickness. Control of stand 5 in this manner is included in this example as being typical of certain installations.

The thickness measurement just after stand 1 is used to control the thickness by adjusting the roll gap actuator at stand 1. However, there are excursions remaining in the thickness that are eventually seen at the downstream stands 2 and 3. It is assumed that there is an operational tracking system in place to track these thickness errors to these stands and change their speeds accordingly to attempt to offer corrections. To maintain mass flow upstream of these stands during these corrective speed changes, the speeds of the upstream stands must also change in correct proportion. Maintenance of the mass flow in the upstream stands in this manner is necessary as it is essential to the stability of rolling.

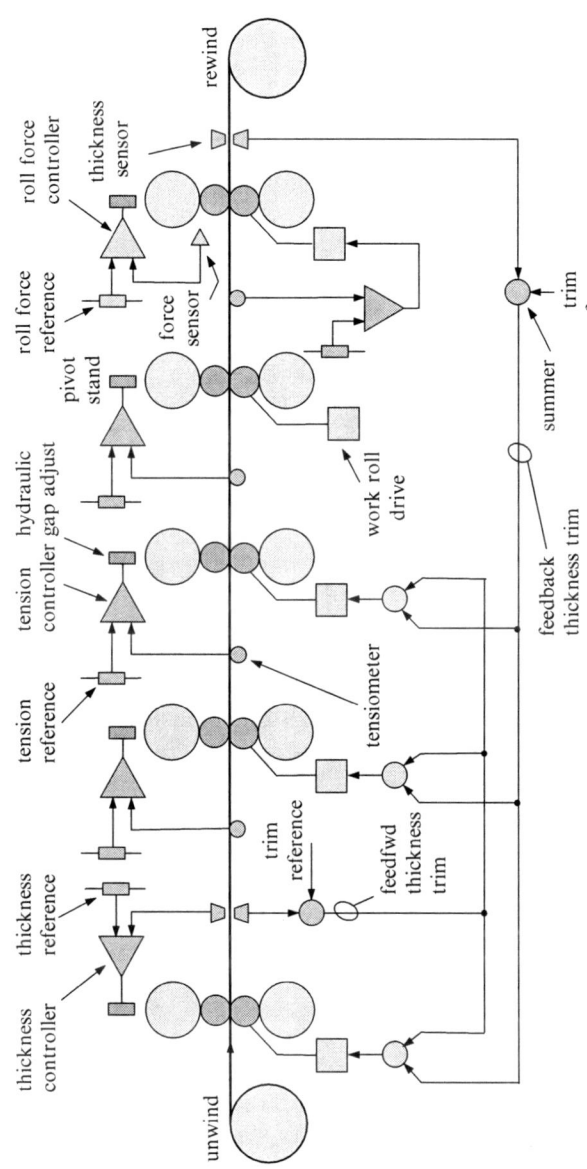

Fig. 3.10 Illustrative tandem cold mill control structure (Based on [7])

This brief analysis of a basic structure of a conventional control strategy for tandem cold rolling illustrates some of the previously noted ideas involved in conventional control techniques, and can be useful as a basis for the understanding of both conventional and more advanced approaches.

3.4.5 Examples of Conventional Control

3.4.5.1 Example 1

Figure 3.11 depicts an example of a conventional control structure, which for ease of understanding is a slightly modified and simplified form of an actual application [8]. The control for this example is close to the illustrative example presented in Section 3.4.4, except with some variations at the entry of the mill plus a few other variations.

As depicted in Figure 3.11, the interstand tensions upstream of stands 2 through 4 are controlled by the "tension by gap" technique. The interstand tension upstream of stand 5 is controlled as "tension by speed," since the roll gap position actuator is used to control the surface finish of the finished product exiting the mill by controlling the roll force.

The control of the thickness at the exit of stand 1 is by feedforward and feedback using thickness measurements at the entry and exit of the stand. The feedforward offers a fast correction to excursions in the incoming thickness. Variations in the exit thickness of stand 1 as seen by the thickness gage at the exit of this stand are corrected by the control loop operating on the roll gap position actuator. These thickness variations for example can be caused by variations in the hardness of the incoming material which would not be seen as thickness variations at the entry thickness gage, but would affect the thickness at the output of stand 1.

The mass flow into stand 2 is controlled to attempt to keep the thickness at the output of stand 2 approximately constant. This done by the mass flow control function, which adjusts proportionally the speed of stand 1 depending on the thickness variation as measured by the thickness gage at the exit of stand 1, so that the thickness at the exit of stand 2 is held nearly constant, assuming a constant width of the material being processed. This is in accordance with the idea of control by the mass flow technique previously described, wherein if the mass flow is held constant and the speed of the stand is held constant, then the thickness at the exit of the stand will be approximately constant.

The thickness measurement at the exit of the mill is used to correct for slowly varying excursions in the exit thickness, the magnitude of which could be fairly significant. These long-term type variations could arise from effects which change very slowly, such as effects caused by slow changes in the hardness of the incoming material, or other conditions which tend to cause a long-term drifting effect in the mill exit thickness. To correct for these effects feedback is provided to adjust the speeds of the appropriate upstream stands. While it is not shown in Figure 3.11,

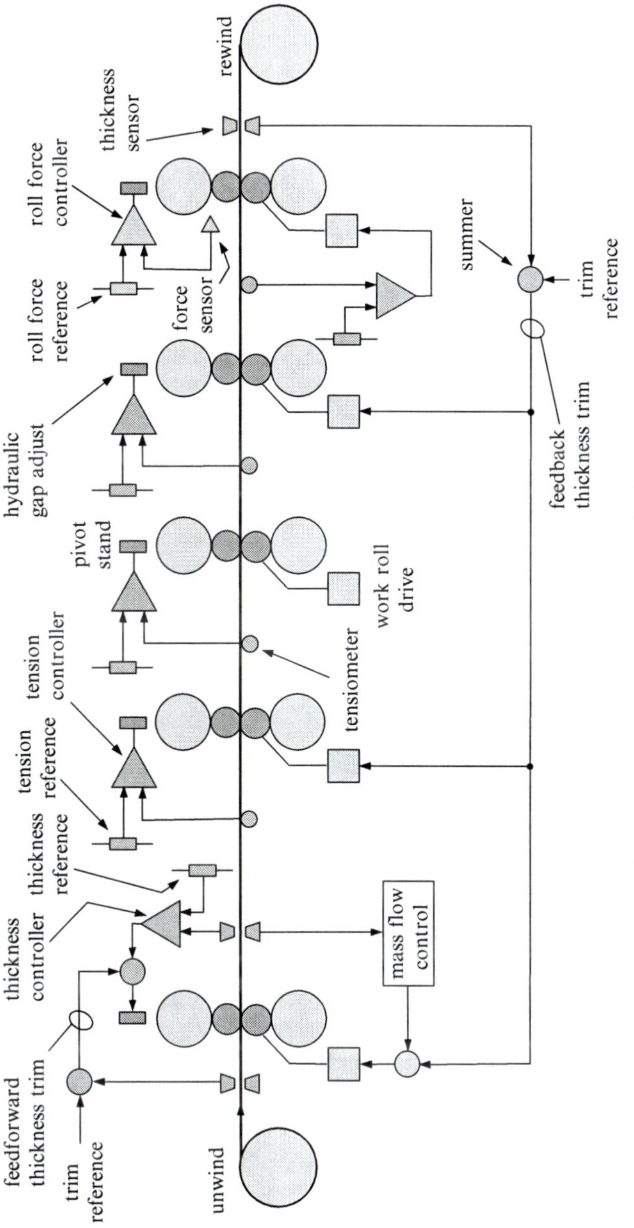

Fig 3.11 Schematic of a typical control structure for a tandem cold mill (Based on [8])

the speeds of these upstream stands are adjusted appropriately to keep the loading of each of the stands within its capability so as to prevent overloading of any of the stands.

Also included but not depicted in Figure 3.11 is the trimming of the strip thickness when there is a tension correction, and the trimming of the tension when there is a strip thickness correction. This reduces the excursions in thicknesses and tensions which contribute to the stability of rolling.

3.4.5.2 Example 2

Figures 3.12–3.14 depict a second example of a method based on conventional control principles, as applied in an actual application [9]. The mill stands in this example are four-high except for stand 5 which is six-high. The intermediate roll in the last stand is used mostly for control of strip flatness which is addressed later in Section 3.6.

This example relies heavily on mass flow techniques to reduce excursions in the strip centerline thicknesses and interstand tensions. One of the objectives of the control structure is to reduce the effects of the delay from the time a change in thickness occurs at the exit of a mill stand until the change is seen at a downstream thickness sensor. This is done by using mass flow techniques to predict the thickness at the exit of the stand. In addition, adaptive learning methods are employed to reduce the error in the mass flow calculations. The estimated thickness $h_{MF,i}$ at the output of stand i as determined by mass flow calculations is

$$h_{MF,i} = A_i \hat{h}_{MF,i}, \tag{3.39}$$

where

$$\hat{h}_{MF,i} = \frac{v_{i-1}}{v_i} h_{m,i-1}(t - \tau), \tag{3.40}$$

with $\hat{h}_{MF,i}$ being the output thickness at stand i prior to the application of a learning algorithm, and where the term A_i represents the output of the learning algorithm that is added to improve the thickness estimate by using the inputs from the thickness measurements and the estimated thicknesses. The terms v_i and v_{i-1} are the measured strip speeds at the entry and exit of stand i, $h_{m,i-1}$ is the strip thickness measured upstream of stand i, τ is the time delay to travel from the thickness gage to the stand, and t is time, with the time-dependency of $h_{m,i-1}$ shown as $h_{m,i-1}(t - \tau)$. Exceptions are in the cases of stands 3 and 5, where there is no upstream thickness measurement. In these instances the thicknesses at the outputs of stands 2 and 4 as estimated by the mass flow computations, are used in place of the measured thickness signals.

As shown in Figure 3.13 the thickness at the output of stand i is controlled by adjusting the roll gap. To reduce the excursions in thickness caused by excursions in the interstand tension, a signal representing the tension error is modified by the

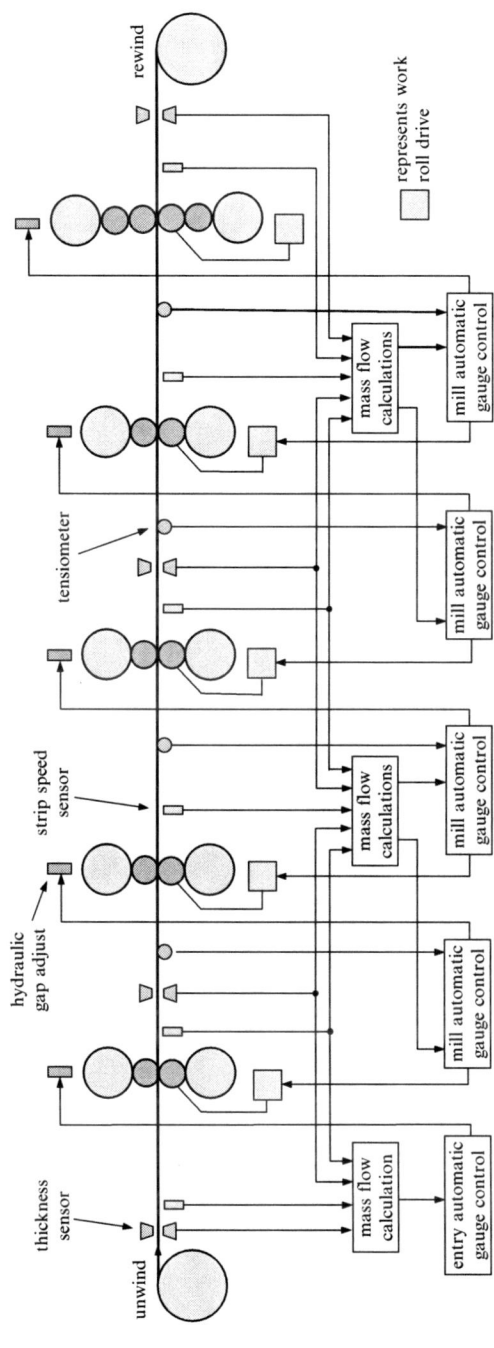

Fig. 3.12 Schematic for conventional control using a mass flow approach (Based on [9])

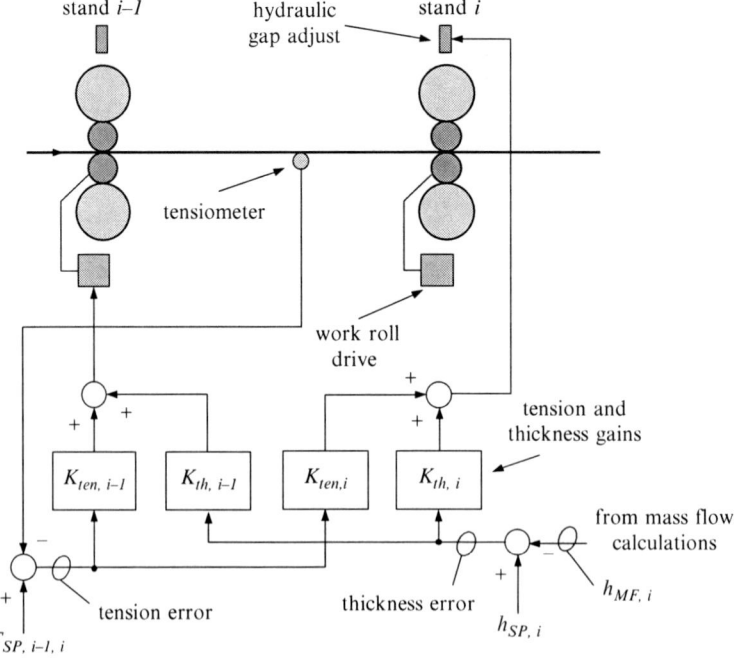

Fig. 3.13 Mill automatic gage control (Based on [9])

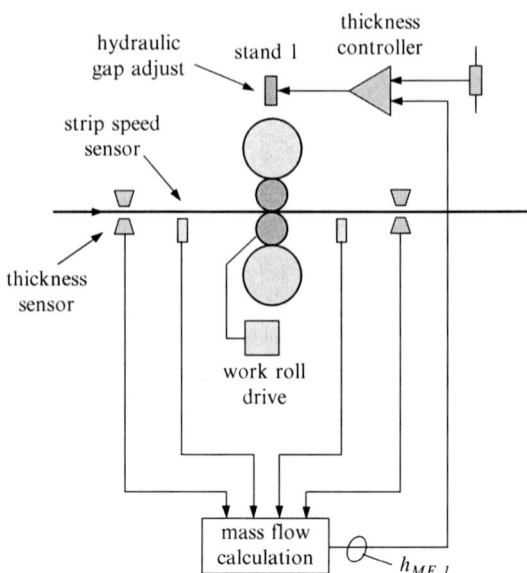

Fig. 3.14 Entry automatic
gage control (Based on [9])

gain term $K_{ten,i}$ and then added to the thickness error signal as modified by the gain term $K_{th,i}$ to reduce excursions in thickness caused by excursions in tension. Also as depicted in Figure 3.13 the interstand tension between stands i and $i - 1$ is controlled by adjustment of the speed of the upstream stand $i - 1$. As in the case of the control of the thickness, a signal representing the thickness error similarly is modified by a gain term $K_{th,i-1}$ and then added to the tension error signal modified by a gain term $K_{ten,i-1}$ to reduce excursions in tension caused by excursions in thickness. The addition of a tension error signal to the thickness error signal, and the addition of a thickness error signal to the tension error signal, is a conventional method of attempting to reduce the interactions between strip thickness and tension. This is important as the stability of rolling depends on the tight control of both thickness and tension not only at the mill output but also at the interstand areas. The control of thickness at the exit of stand 1 (Figure 3.14) is similar to the other stands except that it is unnecessary to include a signal representing the upstream tension which in this case is set by the control of the unwind.

An advantage of this method is that it provides good control of thickness and tension (and thus mass flow) throughout the entire mill, which enhances the stability of rolling and leads to a final product of good quality. However, there are additional sensors for strip speed and thickness beyond what generally is used in other conventional applications, which increases the maintenance burden to retain an acceptable availability of the process. Nonetheless, for a particular installation this could be considered acceptable to achieve a significant improvement in performance.

3.4.6   Conventional Control for Continuous Tandem Mills

Conventional control for continuous tandem cold mills incorporates many of the features presented previously for the conventional control of stand-alone mills. The major additional considerations for control of continuous mills revolve around the change from the present strip being processed to the processing of the next strip, without stopping the mill. Several areas that require consideration when this change occurs (*i.e.* when the weld which joins the two strips moves through the mill) are:

- Modifications to the controller references and settings as the weld moves through each mill stand
- Reduction of out-of-tolerance strip thicknesses in the vicinity of the weld
- Reduction of excursions in the roll force to avoid marking the work rolls
- Maintaining the conservation of mass flow to support the stability of rolling
- Reduction of excursions in interstand tensions
- Tracking of the weld with low error
- Speed reduction as needed to avoid cobbles

How these areas are addressed in certain conventional applications is considered in the material that follows. The intent of what follows is to give a flavor of some of these conventional approaches in response to the above. The presentation is not exhaustive as there are other effective methods which have been implemented successfully. Moreover, it should be noted that many of the methods noted in what follows also are applicable to stand-alone applications where a single larger coil is composed of several strips that are welded together, or where it is sometimes required that the reduction patterns pertaining to a particular coil are deliberately changed during the rolling process.

3.4.6.1 Interfaces with Upstream Processes

Usually in the case of steel the present strip being processed in a continuous mill is joined to the next strip by welding. The weld which joins the two coils of strip is made at the entry portion of the upstream process which is almost always a continuous pickling process. This process feeds the strip continuously into the tandem cold mill through a storage device to accommodate speed changes in the mill, while the speed of the upstream process remains unchanged. Usually this storage device is an accumulator (also denoted as a "loop car" or a "looper"), although in some older installations a loop of strip hanging freely over a pit is used as a storage device to accommodate the differences in speed between the upstream process and the tandem mill. The weld at the entry of the upstream process is made without changing the speed of the process. This is done by using another storage device at the entry portion of the process to accommodate the stopping of this portion of the processs to make the weld.

The strip tension between the upstream process and the tandem cold mill is set by the accumulator, and quite often also with a bridle between the exit of the accumulator and the mill entry. The drive motors of the accumulator and the bridle are controlled to produce the desired tension in the strip. More detail describing the interfaces between the tandem cold mill and the upstream process is given in Section 1.3.2.

3.4.6.2 Welds and Weld Tracking

In general there are two types of welds which join the present coil to the next coil: These are: (1) the butt weld and (2) the laser weld. In the case of the butt weld, the weld is made by bringing the ends of the two strips together under pressure and passing a high current through the interface. In the case of the laser weld, the weld is made by focusing a strong laser beam that melts and evaporates the metal which condenses and then solidifies to join the two strips. After the joint is made, both the laser weld and the butt weld are prepared to produce a smooth thickness change to reduce the likelihood of breakage as the weld travels through the mill. The widths of the laser weld and the butt weld typically are about 4.0 and 8.4 mm respectively [10]. In the case of a thickness transition, the passage of either of these welds through the roll gap results in a smooth but very steep and rapid change between the thickness of the present strip and the thickness of the next strip. This puts requirements on the method of control to reduce undesirable excursions in tension, and to reduce undesirable excursions in roll force which could possibly mark the work rolls.

A reliable system which tracks the weld with low error in the weld position is essential for effective control during the weld passage. For the control methods described herein, it is essential that such a system is operative and that it updates about every 2 ms or less, which is typical for existing hardware and software platforms. The weld tracking system generates signals that indicate the weld position as the weld travels through the mill, and provides a signal to initiate any deceleration needed for the mill to be at constant weld passage speed just before the weld enters the first stand. In addition, as the weld approaches the last stand, the weld tracking system must provide a signal to the mill logic which subsequently sequences the shear and pinch roll and initiates any acceleration needed to go to run speed just after tension is established at the available rewind.

3.4.6.3 Modifications to Controller References During Weld Passage

The product being rolled and the desired mill speed determine an operating point for the mill. During the weld passage, the operating point is changed from the operating point for the present strip to the operating point for the next strip. This is often denoted as a "flying gauge change" and requires that the individual references set in the controller for the roll gap position, tension, and drive speed be changed accordingly as the weld approaches, passes through, and then moves away from the roll gap of each stand. It is desirable that these references are changed in a manner that reduces the length of strip near the weld that has thicknesses that are out-of-tolerance and reduces the excursions in tension and roll force. Further, and quite important, the mass flow balance must be maintained to support the stability of rolling.

While there are many approaches to implementing these changes in the controller references, in general two methods are recognized for changing the references during the weld passage. In the first method, the strip remains in contact with the work rolls during the weld passage and the controller references are adjusted as the weld approaches, passes through, and then exits the roll gap. In the second method, the work rolls are raised slightly to open the roll gap to allow the weld to pass through without contacting the rolls, and then reclose on the strip after the weld has passed. The decision to use either of the two methods generally depends on how severe the transition from one strip to the next is anticipated to be. If the transition is expected to have a weld of high quality and not to have extreme changes in the strip characteristics such as in thickness, hardness, or width, the first method could be used with a resulting shorter length of strip in the vicinity of the weld that has thicknesses that are out-of-tolerance. Otherwise the second method would be used.

Typical sequences for implementing the flying gage change using both of these methods are described in what follows. In both cases a weld tracking system is assumed to be in place and operative to provide signals to the mill logic as needed for initiating actions based on the weld position.

3.4.6.4 Modifications to the Position Actuator References for the First Method

In the first method, as the weld approaches stand 1, a roll gap position actuator movement is initiated to move the actuator from a position corresponding to the characteristics of the present strip to a position corresponding to the characteristics of the next strip. At some point during the movement, the output thickness and roll force will change very rapidly as the weld enters and passes through the roll gap. To reduce the length of strip near the weld that has thicknesses that are out-of-tolerance, the position actuator is moved at nearly its maximum speed (typically about 1.5 mm/s) over a path that reduces both the out-of-tolerance thicknesses and the changes in roll force during the transition. Such a path is one wherein the position actuator movement is initiated to start the transition to the next strip so that the weld passes through the roll gap when the actuator is at about one half of its travel. This allows for some margin around the half-travel point, and thus increases the likelihood that the weld will travel through the roll gap during the movement of the transition to the next strip, which thus reduces the length of strip near the weld that is out-of-tolerance, and reduces the excursions in rolling force that could potentially damage the work rolls. However, if other paths were used such that the weld went through the roll gap before the transition was initiated the excursions in the rolling force could be excessive; if a path was used such that the weld went through the roll gap after initiation of the transition the out-of-tolerance thicknesses could be excessive. Some examples of various paths and their effects on the rolling force and the out-of-tolerance lengths are presented in Chapter 5.

The roll gap position actuator is switched to closed-loop position control during its travel, and after the completion of the travel is returned to its regular mode of control for the next strip, with bumpless transfers of the control modes during the switching. Figure 3.15 depicts the control sequencing for the actuator position during the weld passage. Figure 3.16 shows an example of a thickness transition between the present strip and the next strip, where the weld passes through the roll gap at about the half-travel point, with no change in hardness or width in the next strip or in the weld material.

The sequencing of the position referencing applies primarily to the first stand. To assure that the margin around the half-travel point is retained as the weld moves through the remaining downstream stands, the length of the transition is not decreased as the weld passes through these stands, even though the associated position actuators might be moved at less than their maximum speeds. However

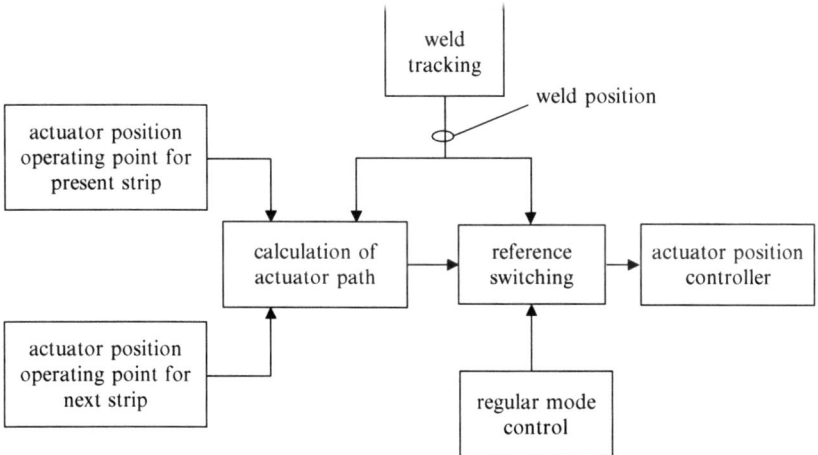

Fig. 3.15 Roll gap actuator position control during a transition from the present strip to the next strip

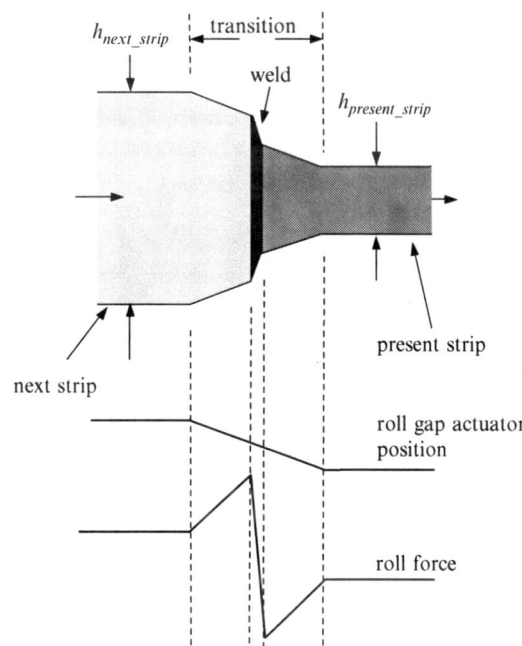

Fig. 3.16 Transition from the present strip to the next strip using the first method

if the length of the weld transition cannot not be retained with a downstream position actuator near its maximum speed, then the length of the transition is increased as determined by the movement of the actuator at its maximum speed.

3.4.6.5 Modifications to the Speed Actuator References for the First Method

Usually transitions between the present strip and the next strip are done when the mill is at a steady speed which in most instances is lower than the run speed, as operation at a reduced steady speed has been found by experience to have less likelihood of strip breakage at the weld. Also, during the transition the mill speed must be reduced as needed to a speed which is compatible with the cutting capabilities of the shear during the transfer to the next available rewind. Further, in many instances the mill speed at transition is lowered to reduce the length of strip near the weld that is out-of-tolerance; some mill speeds during the transition are as low as 5% [11], while others can be as high as 75% or more [12] depending on the weld quality and the severity of the mismatch in the characteristics of the present strip and the next strip.

The speed actuator references are adjusted during the passage of the transition through the mill in accordance with the conservation of mass flow to maintain the stability of rolling. As the transition passes through a mill stand, it is important to coordinate the conservation of mass flow at both the upstream and downstream areas. For example, it can be recognized that during a transition in the thickness at a certain stand, the mass flow at the upstream stands may not be the same as the mass flow at the downstream stands, and that while the thickness transition is passing between one stand and the next, the mass flows at the two adjacent stands are not the same. Thus to conserve mass flow during the transition through the mill, it is necessary that the strip speeds at the exits of the mill stands change as the transition passes from stand to stand. The following example provides an illustration of this.

In this example it is assumed that a five-stand mill is operating at a weld transfer speed of 10% of run speed (120 m/min) at the exit of stand 5, and that there is a 20% increase in the incoming thickness from the present strip to the next strip, with no change in the material hardness or width. Assuming that the exit speed of the mill is to remain unchanged during and after the passage of the transition to the new strip and that there is no change in the strip width during rolling, Table 3.3 lists the strip speeds v_i, the strip thicknesses h_i, and the mass flows MF_i at the mill entry, between mill stands, and at the mill exit using the conservation of mass flow. As can be seen from Table 3.3 the steady-state mass flow changes and is conserved at the upstream and downstream stands as the strip sequences through the mill.

During the passage of the strip transition through the roll gap of stand i, the speed of stand i is held fixed to reduce the likelihood of weld breakage. When the strip transition goes through the roll gap of stand $i + 1$ the speed of stand i, and the speeds of other upstream stands, are coordinated with the changes in the roll gap position of stand $i + 1$ so that the upstream mass flows are conserved during the transition through stand $i + 1$ to support the stability of rolling. At the same time, the mass flows of the stands downstream of stand i retain their steady-state mass flow balances.

Table 3.3 Strip thicknesses, strip speeds, and mass flows during a 20% transition in strip thickness

Variable	Weld position					
	Near mill entry	Between stands 1,2	Between stands 2,3	Between stands 3,4	Between stands 4,5	Past mill exit
h_{in1}	3.56 mm	4.27	4.27	4.27	4.27	4.27
h_{out1}	2.95	3.52	3.52	3.52	3.52	3.52
h_{out2}	2.44	2.44	2.76	2.76	2.76	2.76
h_{out3}	2.01	2.01	2.01	2.26	2.26	2.26
h_{out4}	1.67	1.67	1.67	1.67	1.81	1.81
h_{out5}	1.58	1.58	1.58	1.58	1.58	1.71
v_{in1}	53.3 m/min	53.0	49.8	49.9	48.2	48.1
v_{out1}	64.3	64.3	60.5	60.5^+	58.3	58.2
v_{out2}	77.1	77.1	77.1	77.2	74.4	74.3
v_{out3}	94.3	94.3	94.3	94.3	90.9	90.1
v_{out4}	113.8	113.8	113.8	113.8	113.8	113.5
v_{out5}	120.0	120.0	120.0	120.0	120.0	120.0
MF_{in1}	0.190	0.226	0.213	0.213^+	0.206	0.205
MF_{out1}	0.190	0.226	0.213	0.213^+	0.206	0.205
MF_{out2}	0.190	0.190	0.213	0.213^+	0.206	0.205
MF_{out3}	0.190	0.190	0.190	0.213^+	0.206	0.205
MF_{out4}	0.190	0.190	0.190	0.190	0.206	0.205
MF_{out5}	0.190	0.190	0.190	0.190	0.190	0.205

Dimensions of MFs are m^2/min, *i.e.*, mass flow/width/minute

3.4.6.6 Control of Interstand Tension for the First Method

In those cases where stand i is the next stand through which the weld is to pass and also is controlling the interstand tension between stand i and stand $i - 1$ using the tension by gap approach, as the weld approaches stand i the control of the interstand tension is switched (bumplessly) to tension by speed using stand $i - 1$ to trim the strip speed. After the weld has passed through stand i, the control is switched back to tension by gap using stand i. Where a change in interstand tension is required for the next strip, the reference for interstand tension between stands i and $i + 1$ is changed slowly from the reference for the present strip to the reference for the next strip as the weld passes between stands i and $i + 1$.

The change is initiated after the weld exits the roll bite of stand i and is completed before the weld enters stand $i + 1$.

3.4.6.7 Control for Use of the Second Method

As noted previously, generally the second method is used where the weld is of lower quality, or where the characteristics of the two strips that are joined would make the transition sufficiently severe that the first method cannot be used.

Most of the approaches previously presented for the first method are also applicable in the case of the second method. However, unlike the control of the gap position in the first method, in the control of the gap position for the second method the roll gap is opened and thus unloads the mill stand during the passage of the weld. As the weld approaches a mill stand, a movement of the roll gap position actuator is initiated to move the work rolls at nearly maximum speed so that they are slightly touching the surface of the present strip just before the weld is in the roll gap (*i.e.*, the stand is essentially unloaded with the work rolls just touching the strip). At this point the roll gap is opened to allow the weld to pass through without touching the work rolls. After passage of the weld, the sequence is repeated except in reverse for the work rolls to just contact the surface of the next strip, and then move to load the stand as required by the rolling schedule for the next strip. This sequence is similar to what is described briefly in [12] and is depicted schematically in Figure 3.17.

The interstand tensions on each side of the roll gap at stand i must approximately match when the roll gap is open, so that when the gap is open the controller at stand $i + 1$ or at stand $i - 1$ would then control the tension on either side of stand i using an average (or some other combination) of the measured tensions at both sides of stand i. The references and feedback signals to each of the controllers are appropriately modified so that one controller is the dominant or lead controller, with the other controller having essentially no effect. After the roll gap is reclosed onto the strip, the modifications to the controller references and the feedbacks are removed and regular control is resumed.

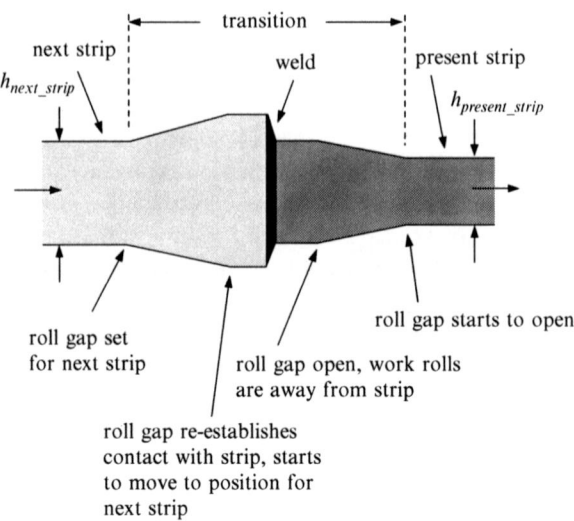

Fig. 3.17 Transition from the present strip to the next strip using the second method (Based on [12])

3.5 Eccentricity Control

As noted previously, roll eccentricity is a general term which refers to any condition caused by axial deviations between the roll barrel and the roll necks that results in irregularities in the mill rolls or the roll bearings. Some examples of these conditions are: (1) eccentricity of the backup roll journals with respect to the roll body, (2) out-of-roundness of the roll, and (3) non-uniformity of rollers in the roller bearings, plus others. These irregularities cause cyclic deviations in the strip thickness at the output of a mill stand. Figure 3.18 depicts the effects of eccentricity on the output strip thickness.

While each of the rolls of the mill stand has some eccentricity which contributes to the cyclic deviation in the strip thickness, the greatest effect is from the backup rolls, mostly because they have a larger diameter than the other rolls. The deviation in strip thickness contributed by the eccentricity in the upper backup roll is not identical to the deviation contributed by the eccentricity in the lower backup roll, because the roll diameters are slightly different. This results in a beat-frequency type phenomenon, wherein the envelope of the higher frequency excursions in the roll force due to the roll eccentricity is modulated by a much lower frequency due to the difference in backup roll diameters.

Depending on the hardness of the strip, eccentricity-type effects can cause imprinting in the strip thickness, and therefore eccentricity effects are undesirable and usually require some method of compensation. This is especially so in the case of the entry stands of the tandem cold mill where the effects of eccentricity are more pronounced due to heavier thicknesses. However, in certain instances, the amount of imprinting is negligible, especially if the hardness of the strip is significantly greater than the structural stiffness of the mill. In addition, the compensation for eccentricity can increase the wear on hydraulic components, so the decision to apply eccentricity compensation in a particular application needs to be carefully weighed.

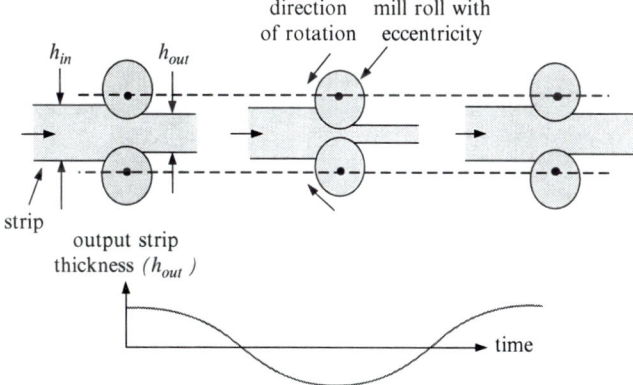

Fig. 3.18 Effects of roll eccentricity on output strip thickness

There are many different methods for compensation of eccentricity. However, in general the type of compensation that is used is one wherein the backup roll angular position is estimated using the measurement of the position of the work roll by sensors (of the pulse tachometer type) used for control of the work roll drives, with some type of calibration to account for errors in the estimates of roll diameters and in changes in roll diameters due to heating and mechanical wear. Based on the measurement of the angular position of the backup rolls the hydraulic cylinder for roll gap positioning is adjusted to compensate for the eccentricity. Calibration of the backup roll position signal often is by an indicator on the backup rolls to recalibrate at each revolution of the roll. Other somewhat more advanced methods use mathematical algorithms to eliminate the indicators on the backup rolls as these devices are subject to higher failure rates and require considerable maintenance to retain their integrity.

An example of a technique [9] of eccentricity compensation using conventional methods is applied to the first stand in a tandem cold mill. In this method the rotational angle of the backup roll is estimated using the input from the pulse tachometer on the work roll with a proximity switch for recalibration at each revolution of the backup roll to determine the backup roll angular position. The idea of this method is to correlate a position on the backup roll with a position on the strip, and then track the position on the strip to the thickness gage located just after the exit of the first stand. At the thickness gage the error in thickness is determined. Then, during the next revolution of the backup roll, when the backup roll is at the same angular position as when the initial error in thickness was determined, a correction value is given to the roll gap position controller to approximately correct for the initial error.

Somewhat more advanced methods use online methods to identify how the roll profiles change with heating and mechanical wear and adjust the roll gap position controller accordingly.

3.6 Automatic Flatness Control

As inferred previously, the control of a tandem cold rolling process consists of two basic areas, *i.e.*, (1) the control of strip thickness and tension which come under the heading of automatic gage control (AGC) that has been addressed in the foregoing, and (2) automatic flatness control (AFC) which will be considered briefly in this section. Generally these two areas can be considered separately as the interactions between them usually are less significant. In this section, this will be assumed to be true so that the automatic flatness control can be addressed apart from the automatic gage control.

While there is no formal definition of strip flatness or strip shape, for intuitive purposes flatness can be understood as the ability of the strip to lie flat on a flat surface under the influence of only gravity. Another and somewhat more formal definition of flatness is the amount of difference in the internal stress across the width of the material, or more specifically the geometric departure from a reference plane when the strip is not under tension and at ambient temperature [13]. For the purposes of this work, strip flatness and strip shape are considered to represent the same strip characteristics.

3.6.1   Causes of and Measurement of Deviations in Strip Flatness

It would desirable if strips processed in the hot rolling area could be of the same thickness across their widths and along their lengths, and also be free of any residual stresses that might cause distortions in their flatness. However, it is not practical to produce such hot-rolled product for a variety of reasons, which are as noted for example in Roberts [14]. When such strips are processed in a tandem cold rolling mill, their thickness profile must remain the same as when leaving the hot mill, except for some slight areas at the strip edges. If there is an attempt (whether intentional or unintentional) to change the thickness profile during cold rolling, additional residual stresses in the strip could be created which lead to distortions (*i.e.* defects) in the flatness of the strip. An example of this is non-uniform strain when the percentage reduction is not uniform across the width of the strip. This could result when, across the strip width, some lengths of the strip come out of the roll bite at different speeds than other lengths, with the resulting defects in flatness. Typical common observable patterns of such defects in the flatness that can be caused by residual stresses are depicted in Figure 3.19. Some additional observable patterns are given in Roberts [5].

To get some further feel of how these defects in the strip flatness can occur, it is helpful to consider the strip as being made up of many narrow strands along its length. If the internal stresses are uniform across the strip width, the strip will be flat in accordance with the above definition. In this case the lengths of the narrow strands of strip will all be of the same length. However, if the internal stresses acting across the strip width are sufficiently non-uniform to overcome the strip's latent internal forces, the narrow strands instead of being of the same length are now of different lengths with the resulting observed deviations in the flatness. It also can occur that a strip may appear to be flat visually according to the above definition; however there may still be non-uniform internal stresses that are hidden to an observer's vision, but become apparent when the strip is slit, sheared, or otherwise worked, to reduce the internal forces that made the strip appear to be flat prior to being worked, with the resulting visual distortion in the flatness after being worked.

Figure 3.20 depicts an ideal condition related to the measurement of flatness wherein the strands of strip are all about the same length. Under this condition there is perfect flatness. However, due to the non-uniform internal stress patterns across the strip width, the individual lengths can change which indicates a deviation in the flatness. When under uniform tension force in the strip, this difference in length results in a variation in tension stress across the width of the strip which can be sensed by appropriate measuring techniques.

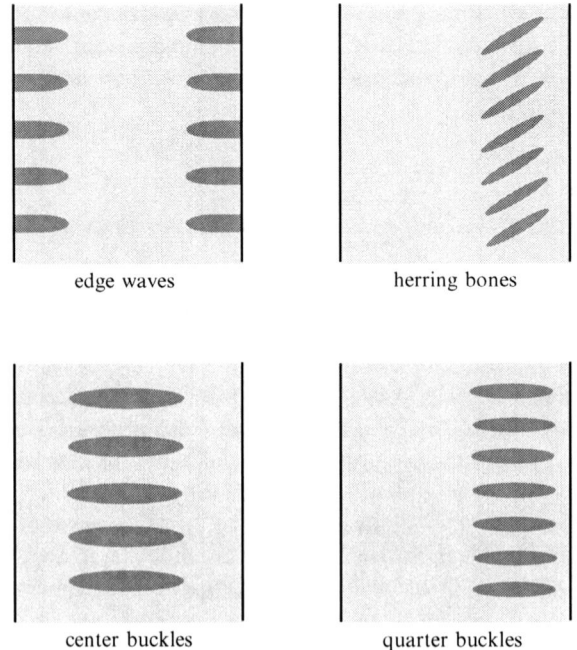

Fig. 3.19 Common strip flatness defects

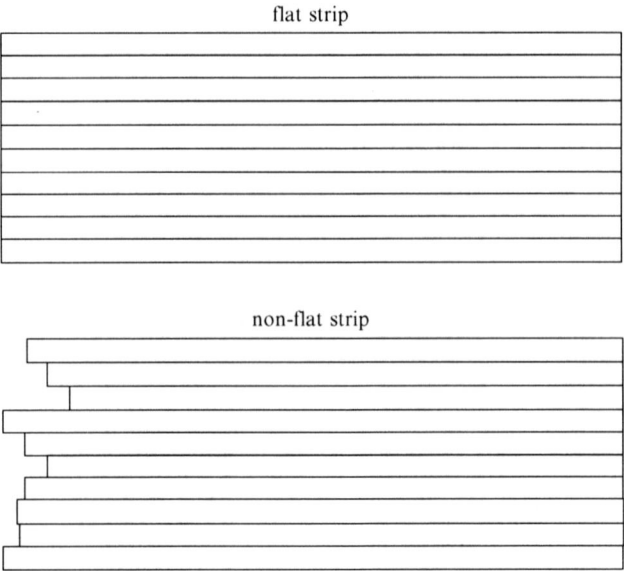

Fig. 3.20 Flatness deviation indicated by differences in lengths of strands of strip

While there are different conventional approaches to the measurement of flatness in rolling mill applications, a device that has seen very wide and successful usage for many years in a great many tandem cold mill applications is the contact shapemeter for cold mills, which is known commercially as a Stressometer. The method used in this device involves the measurement of the pressure across the width of the strip by using a segmented deflector roll, *i.e.* a measuring roll. The basic idea is that each segment of the measuring roll includes a sensor that measures force so that the variation in the tension stress across the strip width can be inferred from the force measurements along the length of the measuring roll.

To compute the tension stress pattern across the measuring roll the mean value of the measured forces is determined. Using the known values of the strip tension force, the strip thickness, and the strip width, a mean value of tension stress σ_m is computed as

$$\sigma_m = \frac{T}{W h}, \tag{3.41}$$

where T is the measured strip tension force, W is the strip width and h is the strip thickness. The tension stress σ_k at each segment of the measuring roll is then determined as

$$\sigma_k = \sigma_m \frac{F_k}{F_m}, \tag{3.42}$$

where F_k is the measured force at roll segment k and F_m is the mean of the measured forces at the roll segments influenced by the strip. The deviations in the individual tension stresses from the mean tension stress are computed as

$$\delta\sigma_k = \sigma_k - \sigma_m, \tag{3.43}$$

where $\delta\sigma_k$ is the deviation in tension stress at roll segment k. The tension stress patterns across the strip width then can be translated into differences in elongations of the strip across the direction of rolling.

In almost all tandem mills the Stressometer is mounted between the last stand and the rewind, as this is an area of the mill where sufficient space is available and where the strip tension is usually well-controlled as is needed for the successful winding of an exit coil. In addition, the cost of Stressometers is quite high which further precludes their multiple use in tandem cold rolling applications.

Figure 3.21 shows a schematic of a typical measuring roll or Stressometer which can have 31 segments, or measuring zones, for a maximum strip width of 1,550 mm. Measurement response times to a step change in force at the Stressometer are on the order of 5 ms. A visual readout derived from measurements by the Stressometer is depicted in Figure 3.22.

Other conventional approaches to the measurement of strip flatness are in use and are applicable to various other applications, such as the need for measurement

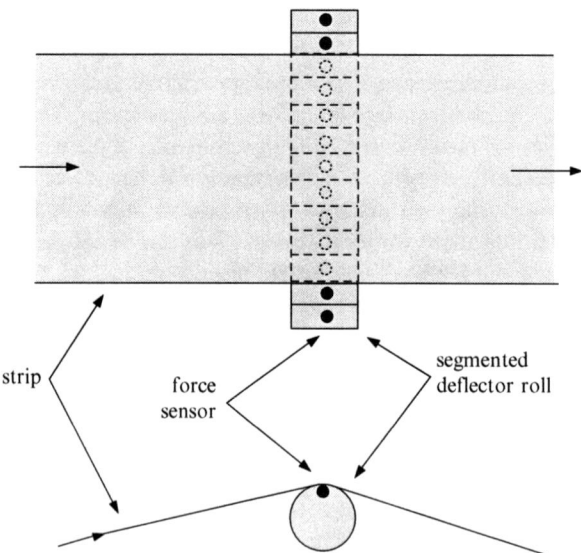

Fig. 3.21 Schematic of Stressometer for tandem cold mills

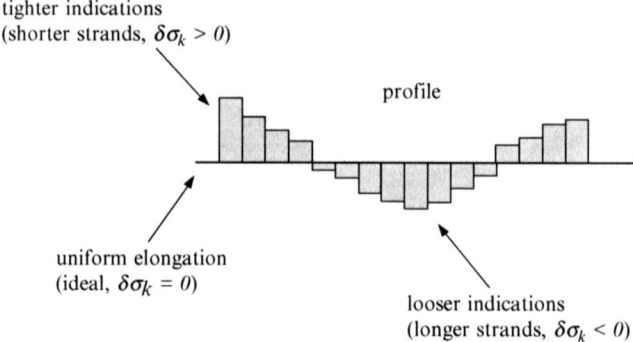

Fig. 3.22 Display based on Stressometer output

without contacting the strip. A brief summary of these other techniques is provided in [13, 15].

3.6.2 Actuators for Control of Strip Flatness

The actuators for flatness control are associated with two types of mill stand configurations: (1) the six-high stand, and (2) the four-high stand. Figure 3.12

depicts both types of these configurations. Where the last stand in the mill is six-high, it is used as the dominant control for strip flatness as it can correct for a wider range of flatness defects than the four-high configuration, and also it is nearest to the flatness measuring roll; however, the other stands also are involved in the flatness control. For purposes of this work it will be considered that the mill configuration consists of four-high upstream stands with the last stand being a six-high configuration, although the concepts presented also could apply to a four-high configuration. A schematic of the six-high stand and the actuators associated with flatness control is shown in Figure 3.23.

The actuator functions are as follows:

- Roll bending puts a slight concave or convex curvature to the roll, while differential roll bending (work rolls only) puts a different roll bending between the two sides of the mill
- Roll shifting shifts the associated rolls laterally with respect to each other
- Roll crossing (intermediate rolls only) shifts the intermediate roll with reference to the work roll, resulting in a curvature of the work roll
- Work roll cooling affects the thermal crown of the roll
- Roll tilting (not shown) causes a difference in the roll gap opening on each side of the stand

Each of the actuators has an effect on the tension stress distribution across the strip width. The roll bending actuators are fast responding, while the intermediate roll crossing actuator is much slower and therefore usually it is set initially based on set-up calculations and is seldom adjusted during rolling. The work roll shifting is set based on set-up calculations and is used mostly to control the thickness of the strip at its edges. It is not adjusted during rolling. Some of the actuators, such as the roll bending actuators, have a limited operating range and therefore are subject to saturation during their control movements. In cases where an actuator is saturated or is close to saturation, another actuator often can be adjusted slightly to avoid the saturation or to pull out of it. An example is the adjustment of the slower intermediate roll crossing actuator to move the faster work roll bending actuators or the

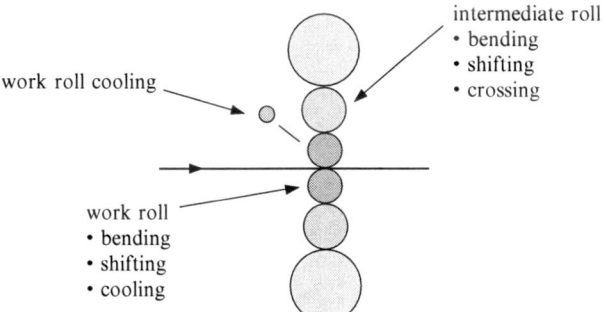

Fig. 3.23 Schematic of six-high mill stand showing flatness-associated actuators

intermediate roll bending actuators away from the saturation limit. The work roll cooling function, which also is slow compared to the bending actuators, uses an array of water sprays to change the thermal crown of the work rolls. Each actuator has a dynamic response which must be considered in the design of the flatness controller. In the case of the four-high mill the actuators are those which implement the work roll bending, tilting, and the work roll cooling functions.

In general the roll bending functions are used to correct flatness defects which are symmetric about the strip centerline, and sometimes less severe defects which are not symmetric. The differential roll bending is used for defects which occur near the edges of the strip, while the roll tilting is used to correct more severe non-symmetric defects in the flatness.

3.6.3   Process Models for Control of Strip Flatness

Mathematical models of the process that are used in the control of strip flatness fall into two categories: (1) models used for the initial setting of the actuators based on the data for the upcoming product to be processed, and (2) a model used in the closed-loop control during the actual rolling process.

3.6.3.1   Models Used for Initial Set-up of the Actuators

The initial set-ups of the actuators are an important part of the control of the strip flatness. The idea of these set-ups, which work in conjunction with the automatic flatness controller, is to deliberately roll into the strip specific transverse stress patterns such that after rolling the strip will exhibit the desired flatness. The desired traverse stress patterns are described in the form of a target function at each of the mill stands. The determination of this target function for each stand is essentially an art based on experience considering the strip thickness profile, roll profiles and wear behavior, thermal characteristics of the mill and strip, actuator capabilities, plus many other process and equipment characteristics. The measurement-based transverse stress distribution at the exit of the last stand is compared against that generated by the target function for the final product to produce a tension stress difference to be used as feedback for closed-loop control.

Generally there are three types of models that are used to establish the initial settings of the actuators to come close to achieving the transverse stress patterns of the target functions. These are: (1) a work roll thermal crown model, (2) a roll stack deformation model, and (3) an internal stress model. The general functions of each of these models are presented briefly in what follows. More specific details related to model development can be found in [16, 17], the references cited therein, and in other associated literature.

The work roll thermal crown model is a dynamic estimation of the thermal expansion (*i.e.* the thermal crown) of the work rolls. This model calculates a temperature field and a profile of the work rolls due to thermal expansion.

The roll stack deformation model is a static model that computes the deformation of the rolls using the rolling force and the thermal crown as determined by the thermal crown model. It also estimates the pressure distribution between the surfaces of the rolls.

The internal stress model is a static model that estimates the internal stresses of the strip resulting from the pressure distribution at the surface of the work roll that is in contact with the strip using the roll deformations as determined by the roll stack deformation model. A profile of the strip also can be generated by this model. Additionally, the profile includes an estimate of the thickness change near the strip edges for correction by the controller as needed.

3.6.3.2 Model Used During Closed-loop Control

The above models are highly complex and their use for closed-loop control during rolling would require that the controller have computational capabilities that are beyond what is presently available in conventional industrial hardware and software platforms. Therefore the model that is used is a simplified and linearized version of the more complex models. The model is applicable at a specific operating point for the present coil and is recomputed for each new coil at its new operating point. The model is in the form of a matrix which maps the position of each actuator to the tension stresses at the exit of the last mill stand under steady-state conditions. For this linearized model, the principle of superposition is assumed to be applicable so that the tension stresses at a particular portion of the strip at the exit of the last mill stand are taken to result from the sum of the stresses caused by the effects of the individual actuators on that portion of the strip.

For example for an actuator j at a particular stand the effect on the tension stresses can be represented as a column vector $\delta\sigma \in R^m$, where m is the number of tension stresses measured at the Stressometer, and the change in the individual stresses is represented as

$$[\delta\sigma_1 \ \delta\sigma_2 \ ... \ \delta\sigma_m]' = Q_j \delta A_j, \tag{3.44}$$

where $\delta\sigma_k$ is the change in the tension stress at a position across the strip width at the exit of the last mill stand that corresponds to zone k of the Stressometer, δA_j is the change in the position of actuator A_j, $Q_j \in R^m$ is a column vector that maps the change in the position of A_j to $\delta\sigma$, and ′ indicates the transpose of a vector or matrix.′ Assuming that there are n actuators in the entire mill, then the relationship that maps the steady-state change in the stresses to the changes in the actuators is

$$[\delta\sigma_1 \ \delta\sigma_2 \ ... \ \delta\sigma_m]' = Q \delta A, \tag{3.45}$$

where $Q \in R^{m \times n}$ is a matrix that is a collection of n column vectors for the n actuators, and $\delta A \in R^n$ is a vector representing the changes in the actuators in the entire mill. The matrix Q is static, *i.e.* its elements do not change during the rolling of a coil.

The dynamics of the process include speed-dependent time delays of the form $e^{-\tau_{i,i+1} s}$ where $\tau_{i,i+1}$ is the delay time for a section of the strip to move between mill stands i and $i + 1$, which is used for tracking purposes. In addition, an expression $G(s)$ for the movement of a section of the strip from the exit of the roll bite of the last stand to the Stressometer is

$$G(s) = \frac{e^{-\tau_1 s}}{(1 + \tau_2 s)}, \qquad (3.46)$$

where τ_1 is the speed-dependent time delay from the exit of the last stand to the Stressometer, and τ_2 is a speed-dependent time constant. The time constant τ_2 arises from St. Venant's principle [6] which accounts for the exponential decay toward zero of the stresses in the strip between the exit of the roll bite and the coiler, so that there is some decay of the stresses between the exit of the stand and the Stressometer. The speed-dependent time delays τ, τ_1, and τ_2 are all of the form distance/strip speed. The dynamics of the faster acting actuators are taken to a first approximation to be single first order lags with a time constant on the order of 100 ms. The response to a step change in each of the forces measured at the Stressometer is as noted previously.

In what follows in Section 3.6.4, appropriate portions of the model are used as needed to make the necessary predictions to compute the changes in settings of certain actuators that are adjusted during actual rolling.

3.6.4 Flatness Controller

While there are different approaches for flatness control a concept that depicts some of the ideas of a basic approach, somewhat similar to [18], is presented in what follows. In this method of control it is assumed for generality that certain actuators of more than one stand can be utilized during actual rolling to correct for the defects in flatness. In many instances only actuators of the last stand are utilized during rolling as this stand is nearest the Stressometer and has a more immediate effect on the flatness; however in other applications certain actuators of some of the remaining stands also may be used in an automatic mode of control. In this method for the control of flatness, the references for actuators used during rolling at a given stand i are determined as depicted in the block diagram of Figure 3.24. The block diagram is a simplified representation to briefly highlight the salient features of this method. This method assumes that the actuators for each stand have been initially set to positions determined by the desired target function.

The idea of the technique depicted in Figure 3.24 is to adjust the actuator references for stand i for a certain length of strip, denoted as length L, as it progresses through the mill, and then repeat the process for succeeding lengths of strip that pass through stand i. The steps in this method are as follows:

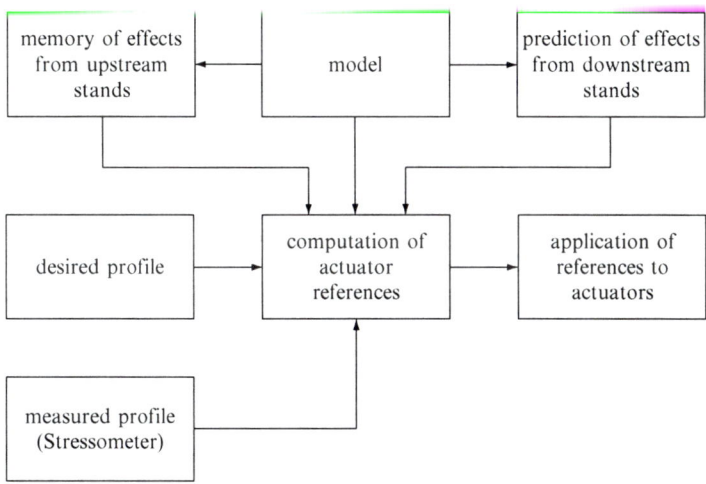

Fig. 3.24 Simplified block diagram of a controller for actuators for stand i

- The effects on the length L of strip when it had passed through the roll gaps of stand 1 to stand $i - 1$ are determined using the model and the memorized settings of the actuators of these upstream stands
- Similarly, the predicted effects on length L of the strip when it will pass through the roll gaps of stand $i + 1$ to stand n, where n is the last stand, are determined using the model and the predicted settings of the actuators for these downstream stands. These predicted settings are the settings of the actuators of the previous length of strip (*i.e.* length $L - 1$) that had passed through these downstream stands. Where the length $L - 1$ had not yet passed through a downstream stand, the settings for length $L - 2$ are used for that stand. Where length $L - 2$ had not yet passed through a downsteam stand, length $L - 3$ is used, and similarly for length $L - 3$ and previous lengths as applicable
- Using the above, the model, the desired target profile determined from the target function, and the actual profile as determined from Stressometer measurements, the estimated effects on the profile at the mill exit are determined, and the necessary changes to the actuator references are computed for the length L of strip at stand i using an appropriate optimization routine. The computed changes to the references then are applied to the actuators for stand i
- The above are repeated for lengths of strip $L + 1$ and succeeding lengths that pass through stand i
- Where stand i is the last stand, the prediction of the effect of the downstream stands is not used. Where stand i is the first stand, the estimate of the effect of the upstream stands is not used

It should be noted that in the above description only the roll bending actuators are utilized for making immediate corrections as these are fast responding, although the slower intermediate roll crossing actuator on the final stand can be used as needed to preclude the saturation of the roll bending actuators. The work roll cooling sprays are used to make slower adjustments to the thermal crown of the rolls.

Additional material relating to various other methods of automatic flatness control can be found in [19–21] and the references cited therein.

3.7 Threading of the Mill

The threading of the mill is a highly important function, particularly in the case of a stand-alone configuration wherein the mill must be successfully threaded to allow the processing of each coil. Therefore some of the basics with respect to the threading and control of the mill are presented in this section, with the aim of highlighting the significant aspects of threading and its control. As a prelude to the presentation of threading, some related definitions are appropriate. These are: (1) bite, (2) stick, and (3) thread. The examples which follow are for the first three stands and are similarly applicable to the remaining stands of the mill.

- Bite is a phase of the threading operation which begins when the strip is moving with its head end just about to enter the roll bite of stand 1, and ends when the head end of the strip is completely inserted into and just past the work rolls so that entry tension and roll force are established at stand 1. The head end of the strip just starts to travel toward stand 2, but is not yet inserted into stand 2. Figure 3.25 depicts the bite phase at mill stand 1. During this phase of threading the strip in stand 1 is being rolled with only entry tension.
- Stick is the next phase after bite. The stick phase starts when the head end of the moving strip is just past stand 1, entry tension and roll force are established at stand 1, and the head end of the strip is moving toward stand 2. The stick phase for stand 1 ends when the head end of the strip is just past stand 2 and interstand

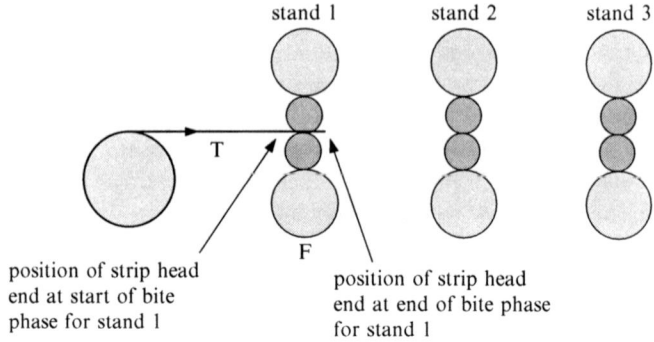

T denotes tension established in area indicated

F denotes roll force established at stand indicated

Fig. 3.25 Mill stand 1 in bite phase

tension is established between stands 1 and 2. At this point stand 2 now is in the bite phase. Figure 3.26 depicts stand 1 in the stick phase and stand 2 in the bite phase. In the case of the last stand in the mill, the stick phase for that stand is concluded when exit tension is established at the rewind.

- The thread phase for stand 1 begins when the head end of the strip is just out of stand 2 and moving toward stand 3, with tension established at the entry to stand 1 and interstand tension established between stands 1 and 2. In this phase the closed-loop controllers for strip thickness, mill entry tension, and interstand tension between stands 1 and 2 are enabled and operational. Stand 1 remains in the thread phase until tension is established at the rewind and the acceleration of the mill to run speed is initiated. Figure 3.27 depicts stand 1 in the thread phase, stand 2 in the stick phase and stand 3 in the bite phase.

3.7.1 Threading Speeds

In the usual and preferred method of threading, each of the mill stands bites at a speed which produces approximately the same steady-state speed at the output of the stand. This requires that the work roll peripheral steady-state speeds be suitably adjusted as the head end of the strip sequences through the mill, which is necessary to maintain the conservation of mass flow to assure the stability of rolling during threading. In addition, this method of threading is friendlier to operational and commissioning personnel, as the head end strip speeds remain essentially unchanged as the head end moves from stand to stand.

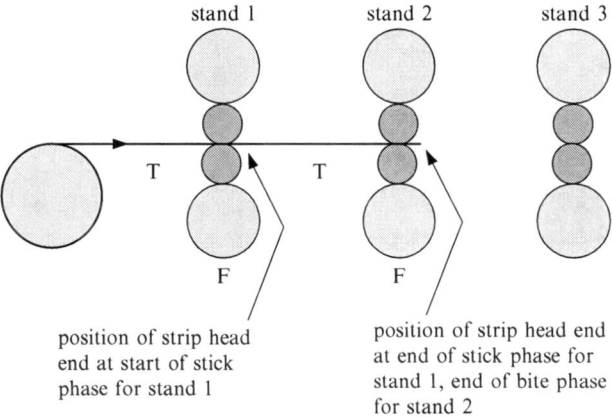

T denotes tension established in area indicated

F denotes roll force established at stand indicated

Fig. 3.26 Mill stand 1 in stick phase, mill stand 2 in bite phase

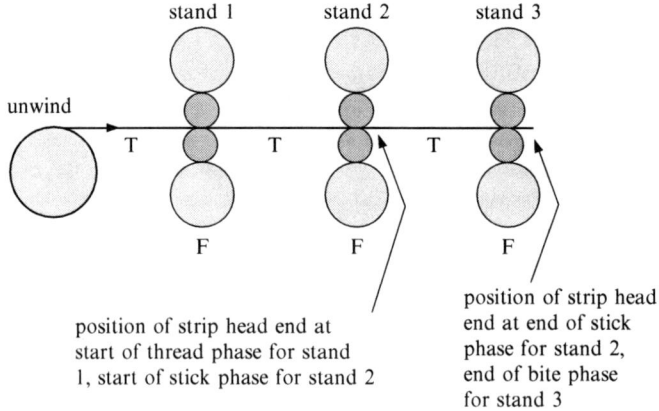

position of strip head end at
start of thread phase for stand
1, start of stick phase for stand 2

position of strip head
end at end of stick
phase for stand 2,
end of bite phase
for stand 3

T denotes tension established in area indicated

F denotes roll force established at stand indicated

Fig. 3.27 Mill stand 1 in thread phase, mill stand 2 in stick phase, mill stand 3 in bite phase

Table 3.4 Work roll peripheral steady-state speeds during threading

Stand in bite phase	Work roll peripheral speed of stand				
	1	2	3	4	5
1.	28.7 m/min	–	–	–	–
2.	22.2	29.4	–	–	–
3.	17.4	23.6	29.5	–	–
4.	13.8	18.7	23.5	29.4	–
5.	12.9	17.5	22.0	27.6	29.9

As an example, if the desired thread speed at the exit of the last stand of a five stand mill is 30 m/min, the steady-state peripheral speed of the work rolls for a certain application is about 29.9 m/min considering the forward slip and the reduction of stand 5. Taking into account the forward slips, the reduction patterns of the mill stands and with constant strip width, a steady-state speed reference for each of the stands can be estimated as shown in Table 3.4, which is based on Production Schedule 2 of Table 2.2. The steady-state speed reference given in Table 3.4 is increased slightly to account for the drop in speed (often denoted as "impact drop") as the head end of the strip enters the mill stand, and is subsequently reduced to the value shown as the head end of the strip goes through the roll bite. This is done to keep from throwing a loop upstream of the stand as discussed further in Section 3.7.2.

In Table 3.4 it is assumed that the tail of the previous strip has left the mill and therefore the downstream speeds are not shown. During the stick phase of stand i, the speeds of stands 1 to i are reduced to the listed speeds corresponding to the bite phase of stand $i + 1$, with a lead speed at only stand $i + 1$ to account for the drop in speed as the head of the strip enters stand $i + 1$. The forward slips assumed in Table 3.4 for each of the stands are 0.044, 0.021, 0.017, 0.019, and 0.004 for stand 1 through stand 5 respectively.

3.7.2 Control During Threading

A significant objective of the control during threading is to reduce the length of strip whose thickness is out-of-tolerance. The idea of various conventional methods is to enable a thickness control function as soon as interstand tension is established between two adjacent mill stands and thus reduce the length of strip whose thickness is not within acceptable limits. Generally it is common practice for thickness measurements to be located at the exit of stand 1 and at the exit of the last stand in the mill which is taken to be stand 5. The thickness measurement at the exit of stand 1 is used to provide feedback to a thickness controller which adjusts the work roll gap actuator of stand 1. This is assumed in the following description of the sequence of events for the control of strip thickness and tension as the strip is threaded through the mill.

- The strip is introduced into stand 1 so that stand 1 is in the bite phase and then in the stick phase as strip moves toward stand 2. The strip then enters stand 2 which is first in the bite phase and then in the stick phase.
- When stand 2 enters the stick phase, the controller for control of thickness at the exit of stand 1 is activated, and the control of interstand tension by speed control of the work rolls of either stand 1 or stand 2 is enabled.
- When stand 3 enters the bite phase, the thickness out of stand 2 is controlled, and the control of interstand tension between stands 1 and 2 is switched to control by adjustment of the roll gap of stand 2. Any feedforward control of thickness also is activated. Control is enabled to adjust the interstand tension between stands 2 and 3 by adjusting the speed of stand 2 or stand 3.
- A similar sequence continues as the head of the strip moves through the mill until the strip is threaded onto the rewind and tension is established between the rewind and stand 5. The thickness controller, using feedback from the thickness measurement at the mill exit, then is activated to control thickness. The method of exit thickness control will depend on the processing requirements of the particular installation.
- The control of interstand tension between stands 4 and 5 likewise will depend on the particular processing requirements, $e.g.$ tension control in this area may be by speed where it is desired that the last stand is controlled to produce constant roll force for setting a particular surface quality of the exiting strip ($e.g.$ Figure 3.10).

As previously noted it is common practice to add a slight lead speed to the mill stand in the bite phase to approximately compensate for the impact drop in the work roll peripheral speed during the bite, so that when the strip enters the stand the work roll speed is close to the desired speed. The biting of the stand causes a nearly step disturbance in the load torque of the work roll speed controller, which results in a rapid decrease in the work roll speed. The lead speed provides some compensation during the impact drop and the subsequent recovery from the disturbance by the speed controller.

3.8 Concluding Comments

This chapter presented several concepts for control of the various areas of the tandem cold rolling process. In general the concepts presented are typical of basic conventional methods for control of the mill. While additional emphasis is placed on the automatic control of thickness and tension, other areas such as the control of flatness and threading are not neglected to provide a more complete picture of conventional mill control. While more detail relating to each of the concepts presented is available in the cited references and other pertinent literature, it is considered that what is presented forms a useful basic background for understanding other conventional concepts and the advanced techniques presented in Chapters 4 and 5.

References

1. Roberts WL. Cold rolling lubrication. In: Cold rolling of steel. New York: Marcel Dekker; 1978.
2. Lenard JG. Tribology. In: Primer on flat rolling. Oxford, UK: Elsevier; 2007.
3. Wood GE, Ivacheff DP. Mill modulus variation and hysteresis-their effect on hot strip mill AGC. In: Iron and steel engineer yearbook. Pittsburgh: Association of Iron and Steel Engineers; 1977.
4. Zeltkalns A, Riciatti RL. Force sensing in rolling mills. Iron and steel engineer yearbook. Pittsburgh: Association of Iron and Steel Engineers; 1977.
5. Roberts WL. Strip shape: its measurement and control. In: Cold rolling of steel. New York: Marcel Dekker; 1978.
6. Bryant GF. The automation of tandem mills. London: The Iron and Steel Institute; 1973.
7. Geddes EJM. Tandem cold rolling and robust multivariable control, Ph.D. thesis. Leicester: University of Leicester; 1998.
8. Duval P, Parks JC, Fellus G. Latest AGC technology installed at LTV's Cleveland 5-stand cold mill. Iron Steel Eng. 1991;69(11):46–51.
9. Tezuka T, et al. Application of a new automatic gauge control system for the tandem cold mill. IEEE Trans Ind Appl. 2002;38(2):553–8.
10. Binroth C, Fedosseev A. Behavior of weld seams during cold rolling processes. Proceedings of the 9th International, 4th European Steel Rolling Conference Association Technique de la Sidérurgie Française; 2006; Paris.
11. Keintzel G, et al. Automation of a linked pickling and tandem mill. Rev Metall. 2001;10:861–71.
12. Tomasic M, Felkl J. Rolling of transitions in a continuous tandem cold mill. In: Proceedings of the 9th International, 4th European Steel Rolling Conference Association Technique de la Sidérurgie Française; 2006; Paris.

13. Zipf M. Innovations in shape measurement and control for cold-rolled flat strip products. In: Ginzburg VB, editor. Flat-rolled steel products. Boca Raton: CRC Press; 2009.
14. Roberts WL. Shape of cold-rolled strip. In: Flat processing of steel. New York: Marcel Dekker; 1988.
15. Miani F, Patrizi P. Fundamentals of online flatness measuring devices. In: Ginzburg VB, editor. Flat-rolled steel products. Boca Raton: CRC Press; 2009.
16. Guo RM. Development of an optimal crown/shape level-2 control model for rolling mills with multiple control devices. IEEE Trans Control Syst Technol. 1998;6(2):172–9.
17. Guo RM. Optimal profile and shape control of flat sheet metal using multiple control devices. IEEE Trans Ind Appl. 1996;32(2):449–57.
18. Bemporad A, Cuzzola FA, et al. Optimization-based automatic flatness control in cold tandem rolling. J Process Control. 2010;20(4):396–407.
19. Ringwood JV. Shape control systems for Sendzimir steel mills. IEEE Trans Control Syst Technol. 2000;8(1):70–86.
20. Grimble MJ, Fotakis J. Design of strip shape control systems for Sendzimir mills. IEEE Trans Automat Control. 1982;27(3):656–66.
21. Hoshino I, et al. Observer-based multivariable features control of a cold rolling mill. Control Eng Pract. 1993;1(6):917–25.

Chapter 4
Advanced Control

4.1 The Need for Advanced Control

The conventional methods for control of the tandem cold metal rolling process generally have proved to be quite successful in producing a product of reasonably good quality. The structure of almost all of the conventional varieties of controllers is usually based on single-input-single-output (SISO) proportional-integral (PI) type control loops which are relatively easy to tune and are user-friendly to design and commissioning personnel, most of whom have limited backgrounds in advanced control theory. In addition, this type of controller structure can be configured to allow controller parameters to be scheduled easily as a function of the mill speed and to deal with the effects of the varying and significant time delays caused by the travel of the strip between the individual mill stands, plus providing a simple means of adjustment to handle changes in the characteristics of the product being processed. All of these features are desirable as they have contributed strongly toward keeping the design and commissioning efforts at reduced levels with the resulting ultimate savings in engineering and production costs.

However, the SISO-type structure of the many conventional methods in use has resulted in major limitations in their ability to improve performance resulting from their inherent limited capability to deal fully with the dynamic interactions of the various process variables, plus the necessity to improve robustness to parameter uncertainties and to internal and external disturbances. Since a highly competitive market keeps increasing its demands for a product with tighter tolerances and other improvements in quality, the need arises to go beyond the limits of the SISO-based design to consider other alternatives such as multi-input-multi-output (MIMO)-based systems as a possible means for product quality improvement, and yet retain many of the desirable features of the conventional methods such as user-friendliness and the simplicity of design and implementation.

In response to this need, several methods of advanced MIMIO control have been proposed and simulated in academia, and some have been implemented in operating installations. In this chapter, we examine two of these methods and compare them to conventional techniques. In addition, Chapter 5 presents a third

J. Pittner and M.A. Simaan, *Tandem Cold Metal Rolling Mill Control*,
Advances in Industrial Control, DOI 10.1007/978-0-85729-067-0_4,
© Springer-Verlag London Limited 2011

technique that offers an excellent potential for improvement in performance and compares the performance and the suitability for usage in actual practice using this technique to conventional methods. Of course not every MIMO method simulated or transformed into actual operation can be addressed. Nevertheless, the methods considered are those that are based on reasonably well-established techniques, have been actually applied or investigated by simulation, and are documented and widely referenced in the literature. The techniques that are examined in this chapter are: (1) a method using an H^∞ loop shaping approach, and (2) an observer-based method for a MIMO controller.

4.2   Linearization of the Process Model

As most of the MIMIO advanced techniques for control of the tandem cold rolling process are based on the use of a linearized model in state-space form, a prerequisite for consideration of these techniques is the development of such a model. The nonlinear model that will be used as a basis for consideration of the process of linearization will be the model presented in Chapter 2. By developing a linearized form of this model at a particular operating point of the nonlinear system, a process model in linear state-space form then can be realized for control system research and development. In the development of the linearized model, it is assumed that the operating point of the nonlinear system is an equilibrium point of the system so that deviations from the operating point result only from control actions or process disturbances. The material in this section is intended to give some feel for the linearization process.

The following equations which are taken or developed from material in Chapter 2 are the basis for describing the tandem cold rolling process. These relationships are those for specific roll force P_i (4.1), specific roll torque G_i (4.2), strip thickness at the stand output $h_{out,i}$ (4.3), interstand tension stress $\sigma_{i,i+1}$ (4.4), strip speed at the stand output $V_{out,i}$ (4.5), the forward slip f_i (4.6), the mass flow across the roll gap MF_i (4.7), the stand input thickness $h_{in,i+1}$ (4.8), and the stand input hardness $k_{in,i+1}$ (4.9), where the subscript i indicates the stand number, with stand i or stands i and $i+1$ to be understood appropriately where no symbol is specifically noted, with other symbols representing the variables or parameters as denoted in Chapter 2, and with the assumption in (4.7) of constant strip width across the roll gap. Additionally, it is assumed the strip tensions at the mill entry and exit are held constant by the coiler controllers. The specific algebraic expressions for (4.1), (4.2), and (4.6) are as presented in Chapter 2.

$$P_i = P_i(h_a, h_{in}, h_{out}, \sigma_{in}, \sigma_{out}, \bar{k}, \mu), \tag{4.1}$$

$$G_i = G_i(h_a, h_{in}, h_{out}, \sigma_{in}, \sigma_{out}, \bar{k}, \mu), \tag{4.2}$$

$$h_{out,i} = S + S_0 + \frac{PW}{M}, \tag{4.3}$$

$$\frac{d(\sigma_{i,i+1})}{dt} \equiv \dot{\sigma}_{i,i+1} = \frac{E(V_{in,i+1} - V_{out,i})}{L_0}, \quad \sigma_{i,i+1}(0) = \sigma_{0,i,i+1}, \tag{4.4}$$

$$V_{out,i} = V(1+f), \tag{4.5}$$

$$f_i = f_i(h_a, h_{in}, h_{out}, \sigma_{in}, \sigma_{out}, \bar{k}, \mu), \tag{4.6}$$

$$MF_i = V_{in}h_{in} = V_{out}h_{out}, \tag{4.7}$$

$$h_{in,i+1}(t) = h_{out,i}(t - \tau_{d,i,i+1}), \tag{4.8}$$

$$k_{in,i+1}(t) = k_{out,i}(t - \tau_{d,i,i+1}), \tag{4.9}$$

A linearized system of equations approximates small excursions around the operating point. For small perturbations in the various variables denoted in (4.1), small changes in the specific roll force are approximated as a linear relationship (4.10),

$$\delta P_i = \xi_{P1,i}\delta h_a + \xi_{P2,i}\delta h_{in,i} + \xi_{P3,i}\delta h_{out,i} + \xi_{P4,i}\delta\sigma_{in,i}$$
$$+ \xi_{P5,i}\delta\sigma_{out,i} + \xi_{P6,i}\delta\bar{k}_i + \xi_{P7,i}\delta\mu_i \tag{4.10}$$

where

$$\xi_{P1,i} = \frac{\partial P_i}{\partial h_a}, \; \xi_{P2,i} = \frac{\partial P_i}{\partial h_{in,i}}, \; \xi_{P3,i} = \frac{\partial P_i}{\partial h_{out,i}}, \; \xi_{P4,i} = \frac{\partial P_i}{\partial \sigma_{in,i}}, \tag{4.11a}$$

and

$$\xi_{P5,i} = \frac{\partial P_i}{\partial \sigma_{out,i}}, \; \xi_{P6,i} = \frac{\partial P_i}{\partial \bar{k}_i}, \; \xi_{P7,i} = \frac{\partial P_i}{\partial u_{u,i}}, \tag{4.11b}$$

and with the partial derivatives (i.e. the linearizing coefficients) evaluated at the operating point and determined by suitable methods as are available for their computation.

Similarly, small perturbations in the specific roll torque (4.2) and the forward slip (4.6) can be approximated respectively as in (4.12) and (4.14),

$$\delta G_i = \xi_{G1,i}\delta h_a + \xi_{G2,i}\delta h_{in,i} + \xi_{G3,i}\delta h_{out,i} + \xi_{G4,i}\delta\sigma_{in,i}$$
$$+ \xi_{G5,i}\delta\sigma_{out,i} + \xi_{G6,i}\delta\bar{k}_i + \xi_{G7,i}\delta\mu_i \tag{4.12}$$

where

$$\xi_{G1,i} = \frac{\partial G_i}{\partial h_a}, \; \xi_{G2,i} = \frac{\partial G_i}{\partial h_{in,i}}, \; \xi_{G3,i} = \frac{\partial G_i}{\partial h_{out,i}}, \; \xi_{G4,i} = \frac{\partial G_i}{\partial \sigma_{in,i}}, \tag{4.13a}$$

and

$$\xi_{G5,i} = \frac{\partial G_i}{\partial \sigma_{out,i}}, \ \xi_{G6,i} = \frac{\partial G_i}{\partial \bar{k}_i}, \ \xi_{G7,i} = \frac{\partial G_i}{\partial u_{u,i}}, \tag{4.13b}$$

$$\delta f_i = \xi_{f1,i}\delta h_a + \xi_{f2,i}\delta h_{in,i} + \xi_{f3,i}\delta h_{out,i} + \xi_{f4,i}\delta \sigma_{in,i}$$
$$+ \xi_{f5,i}\delta \sigma_{out,i} + \xi_{f6,i}\delta \bar{k}_i + \xi_{f7,i}\delta \mu_i \tag{4.14}$$

where

$$\xi_{f1,i} = \frac{\partial f_i}{\partial h_a}, \ \xi_{f2,i} = \frac{\partial f_i}{\partial h_{in,i}}, \ \xi_{f3,i} = \frac{\partial f_i}{\partial h_{out,i}}, \ \xi_{f4,i} = \frac{\partial f_i}{\partial \sigma_{in,i}}, \tag{4.15a}$$

and

$$\xi_{f5,i} = \frac{\partial f_i}{\partial \sigma_{out,i}}, \ \xi_{f6,i} = \frac{\partial f_i}{\partial \bar{k}_i}, \ \xi_{f7,i} = \frac{\partial f_i}{\partial u_{u,i}}, \tag{4.15b}$$

The time delay $\tau_{d,i,i+1}$ in (4.8) and (4.9) that is related to the travel of the strip thickness and the compressive yield stress (hardness) between stands i and $i + 1$ is sometimes approximated by the well-known standard Pade approximation [1] or by a series of first order lags [2]. In either of these instances, a disadvantage is that additional states are added to the state-space realization for the approximation. However an additional disadvantage of the standard Pade approximation is the initial jump to full output. Other methods, such as a modification to the standard Pade approximation [3] by reducing the order of the numerator, or more advanced methods such as those described in [4] and its references also can be considered as possible alternatives. Where modeling and simulation is done using software packages such as MATLAB\Simulink,[1] functions available in these packages can be used effectively for simulation of time delays whose length varies with process conditions. In the standard fourth order Pade approximation the time delay function is approximated as

$$e^{-\tau d s} \cong \frac{1680 - 840(s\tau_d) + 180(s\tau_d)^2 - 20(s\tau_d)^3 + (s\tau_d)^4}{1680 + 840(s\tau_d) + 180(s\tau_d)^2 + 20(s\tau_d)^3 + (s\tau_d)^4}. \tag{4.16}$$

A modified version of this standard Pade estimate is obtained by reducing the order of the numerator

$$e^{-\tau d s} \cong \frac{840 - 360(s\tau_d) + 60(s\tau_d)^2 - (s\tau_d)^3}{840 + 480(s\tau_d) + 120(s\tau_d)^2 + 16(s\tau_d)^3 + (s\tau_d)^4}. \tag{4.17}$$

---

[1]MATLAB and Simulink are registered trademarks of The MathWorks, Inc., Natick, MA 01760-2098.

The approximation using a series of four first order lags has the form

$$e^{-\tau ds} \cong \frac{1}{\left(\frac{\tau_d}{n} s + 1\right)^n},\qquad(4.18)$$

where $n = 4$ for a fourth order approximation. Figure 4.1 displays a comparison of the responses using the above three methods for a unity step input.

While the time delay for excursions in the strip thickness or hardness is significant for travel between adjacent stands, it should be noted that the time delay for an excursion in the tension to travel between adjacent stands is negligible as the tension travels through the strip as the speed of sound.

The result of the linearization of the model is a set of linear ordinary differential equations with constant coefficients that describe the system at the operating point, and which can be put into a linear state space form of a state equation (4.19) and an output Equation (4.20),

$$\frac{dx}{dt} \equiv \dot{x} = Ax + Bu, \quad x(0) = x_0,\qquad(4.19)$$

$$y = Cx + Du,\qquad(4.20)$$

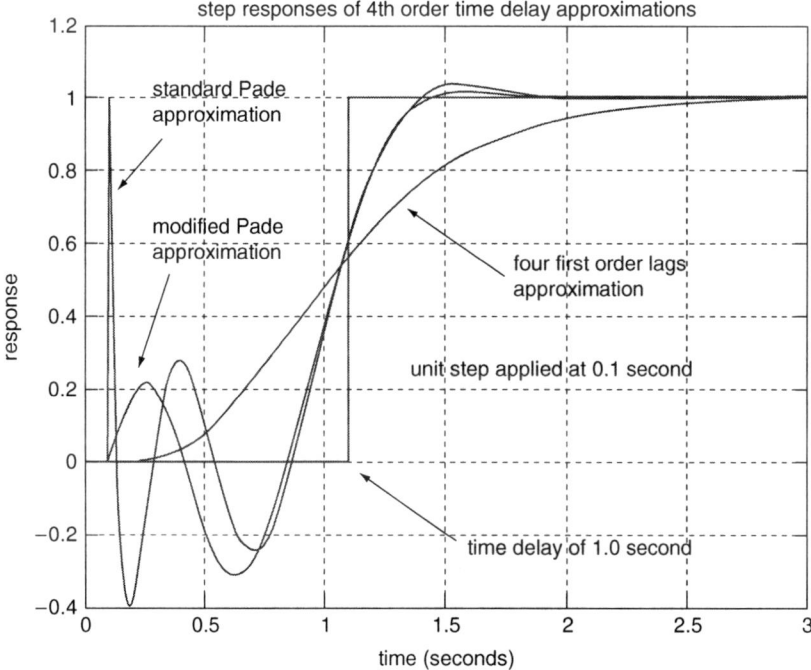

Fig. 4.1 Comparison of responses of time delay approximations

where $x \in R^n$ is the state vector whose elements represent the individual state variables, $y \in R^p$ is a vector whose elements represent the individual output variables, $u \in R^m$ is a vector whose elements represent the individual control variables, and $A \in R^{nxn}$, $B \in R^{nxm}$, $C \in R^{pxn}$ and $D \in R^{pxm}$ are coefficient matrices whose elements are constants that are determined by the linearized equations describing the process. Equations 4.19 and 4.20 can be expanded to include disturbances and uncertainties as needed for analysis of a particular application. It should be noted that the elements of the coefficient matrices, while described above as constants, also could be explicit functions of time and still satisfy the requirements for linearity. More detail related to these requirements can be found in textbooks on linear systems analysis (*e.g.*, [5]). However, if these matrices are modeled as functions of the states, the system might no longer be linear. This will be addressed later in Chapter 5.

As an example to provide more insight into the linearization process, a simple linearized model will be developed for a three stand tandem cold mill at a specific operating point. The mill as depicted in Figure 4.2 can be described by the nonlinear model consisting of Equations 4.1–4.9, except with the understanding that the variables represent small changes from the operating point rather than actual values, so that the δs will be omitted from the equations.

The initial objective is to develop linear differential equations which then can be converted into linear state-space form. For purposes of illustration in this simplified example, only one disturbance will be considered which is a variation in the incoming thickness. It also will be assumed for simplicity that variations in the annealed thickness h_a, the average compressive yield stress (*i.e.*, the average material hardness) $\bar{k}$, and the friction coefficient μ all are zero, that variations in the roll torque G can be neglected, and that there are zero uncertainties in modeling and measurement.

From (4.11a), (4.11b), (4.15a), and (4.15b),

$$P_i = \xi_{P2,i}h_{in,i} + \xi_{P3,i}h_{out,i} + \xi_{P4,i}\sigma_{in,i} + \xi_{P5,i}\sigma_{out,i}, \tag{4.21}$$

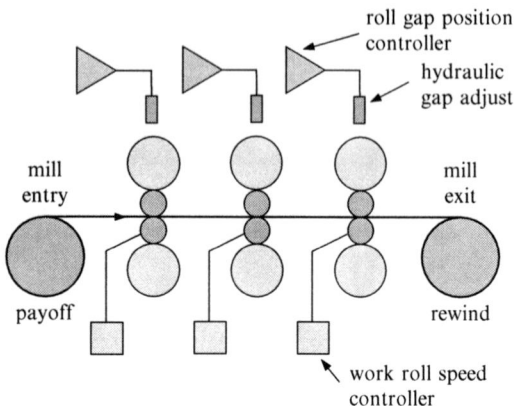

Fig. 4.2 Three-stand tandem cold mill

$$f_i = \xi_{f2,i}h_{in,i} + \xi_{f3,i}h_{out,i} + \xi_{f4,i}\sigma_{in,i} + \xi_{f5,i}\sigma_{out,i}, \tag{4.22}$$

Equations 4.21 and 4.22 can be combined algebraically with (4.3)–(4.7) to give expressions for the variations in the interstand tensions as

$$\frac{d\sigma_{12}}{dt} \equiv \dot{\sigma}_{12} = \zeta_1\sigma_{12} + \zeta_2\sigma_{23} + \zeta_3 h_{in1} + \zeta_4 h_{in2} +$$
$$\zeta_5 S_1 + \zeta_6 S_2 + \zeta_7 V_1 + \zeta_8 V_2, \qquad \sigma_{12}(0) = \sigma_{12,0}, \tag{4.23}$$

$$\frac{d\sigma_{23}}{dt} \equiv \dot{\sigma}_{23} = \zeta_9\sigma_{12} + \zeta_{10}\sigma_{23} + \zeta_{11} h_{in2} + \zeta_{12} h_{in3} + \zeta_{13} S_2 +$$
$$\zeta_{14} S_3 + \zeta_{15} V_2 + \zeta_{16} V_3, \qquad \sigma_{23}(0) = \sigma_{23,0}, \tag{4.24}$$

where the ζs are derived by algebraic manipulation using the ξs, the system parameters, and the steady-state values of the various variables at the operating point.

The closed-loop controllers for the roll gap positions and the work roll peripheral speeds are modeled as single first order lags,

$$\frac{dS_1}{dt} \equiv \dot{S}_1 = \frac{U_{S1}}{\tau_S} - \frac{S_1}{\tau_S}, \quad S_1(0) = S_{10}, \tag{4.25}$$

$$\frac{dS_2}{dt} \equiv \dot{S}_2 = \frac{U_{S2}}{\tau_S} - \frac{S_2}{\tau_S}, \quad S_2(0) = S_{20}, \tag{4.26}$$

$$\frac{dS_3}{dt} \equiv \dot{S}_3 = \frac{U_{S3}}{\tau_S} - \frac{S_3}{\tau_S}, \quad S_3(0) = S_{30}, \tag{4.27}$$

$$\frac{dV_1}{dt} \equiv \dot{V}_1 = \frac{U_{V1}}{\tau_V} - \frac{V_1}{\tau_V}, \quad V_1(0) = V_{10}, \tag{4.28}$$

$$\frac{dV_2}{dt} \equiv \dot{V}_2 = \frac{U_{V2}}{\tau_V} - \frac{V_2}{\tau_V}, \quad V_2(0) = V_{20}, \tag{4.29}$$

$$\frac{dV_3}{dt} \equiv \dot{V}_3 = \frac{U_{V3}}{\tau_V} - \frac{V_3}{\tau_V}, \quad V_3(0) = V_{30}, \tag{4.30}$$

where S, V, τ_S, τ_V, are respectively the roll gap actuator position, the work roll peripheral speed, the roll gap position controller time constant, the work roll speed controller time constant, and U represents a reference variable.

For simplicity in this example, the interstand time delay τ_{12} between stand 1 and stand 2 is approximated as four first order lags in accordance with (4.18) as

$$\frac{dq_1}{dt} \equiv \dot{q}_1 = \frac{h_{out1}}{\frac{1}{4}\tau_{12}} - \frac{q_1}{\frac{1}{4}\tau_{12}}, \quad q_1(0) = q_{10}, \tag{4.31}$$

$$\frac{dq_2}{dt} \equiv \dot{q}_2 = \frac{q_1}{\frac{1}{4}\tau_{12}} - \frac{q_2}{\frac{1}{4}\tau_{12}}, \quad q_2(0) = q_{20}, \tag{4.32}$$

$$\frac{dq_3}{dt} \equiv \dot{q}_3 = \frac{q_2}{\frac{1}{4}\tau_{12}} - \frac{q_3}{\frac{1}{4}\tau_{12}}, \quad q_3(0) = q_{30}, \tag{4.33}$$

$$\frac{dh_{in2}}{dt} \equiv \dot{q}_{hin2} = \frac{q_3}{\frac{1}{4}\tau_{12}} - \frac{h_{in2}}{\frac{1}{4}\tau_{12}}, \quad h_{in2}(0) = h_{in20}, \tag{4.34}$$

The interstand time delay between stands 2 and 3 is approximated similarly as

$$\frac{dr_1}{dt} \equiv \dot{r}_1 = \frac{h_{out2}}{\frac{1}{4}\tau_{23}} - \frac{r_1}{\frac{1}{4}\tau_{23}}, \quad r_1(0) = r_{10}, \tag{4.35}$$

$$\frac{dr_2}{dt} \equiv \dot{r}_2 = \frac{r_1}{\frac{1}{4}\tau_{23}} - \frac{r_2}{\frac{1}{4}\tau_{23}}, \quad r_2(0) = r_{20}, \tag{4.36}$$

$$\frac{dr_3}{dt} \equiv \dot{r}_3 = \frac{r_2}{\frac{1}{4}\tau_{23}} - \frac{r_3}{\frac{1}{4}\tau_{23}}, \quad r_3(0) = r_{30}, \tag{4.37}$$

$$\frac{dh_{in3}}{dt} \equiv \dot{h}_{in3} = \frac{r_3}{\frac{1}{4}\tau_{23}} - \frac{h_{in3}}{\frac{1}{4}\tau_{23}}, \quad h_{in3}(0) = h_{in30}, \tag{4.38}$$

where q_i and r_i are variables related to the approximation of the time delay and do not represent any physical entities.

Similarly for the outputs, algebraic manipulation produces βs for use in Equations 4.39–4.44 for the specific roll forces and the output thickness,

$$P_1 = \beta_1 h_{in,1} + \beta_2 S_1 + \beta_3 \sigma_{12}, \tag{4.39}$$

$$P_2 = \beta_4 h_{in,2} + \beta_5 S_2 + \beta_6 \sigma_{12} + \beta_7 \sigma_{23}, \tag{4.40}$$

$$P_3 = \beta_8 h_{in,3} + \beta_9 S_3 + \beta_{10} \sigma_{23}, \tag{4.41}$$

$$h_{out,1} = \beta_{11} h_{in,1} + \beta_{12} S_1 + \beta_{13} \sigma_{12}, \tag{4.42}$$

$$h_{out,2} = \beta_{14} h_{in,2} + \beta_{15} S_2 + \beta_{16} \sigma_{12} + \beta_{17} \sigma_{23}, \tag{4.43}$$

$$h_{out,3} = \beta_{18} h_{in,3} + \beta_{19} S_3 + \beta_{20} \sigma_{23}. \tag{4.44}$$

The 16 linear differential equations (4.23)–(4.38) and six output equations (4.39)–(4.44) can be put into linear state-space form

$$\frac{dx}{dt} \equiv \dot{x} = Ax + Bu + D_{in}d, \quad x(0) = x_0, \tag{4.45}$$

$$y = Cx + D_{out}d, \tag{4.46}$$

where $x \in R^{16}$, $A \in R^{16 \times 16}$, $B \in R^{16 \times 6}$, $y \in R^8$, $u \in R^6$, $C \in R^{8 \times 16}$, $D_{in} \in R^{16}$, $D_{out} \in R^8$ and for this example the disturbance $d = h_{in1}$ is a scalar. The variables represented by the elements of the state, output and control vectors are listed Table 4.1.

The elements of matrices A, B, C, are as shown in Tables 4.2–4.4. Unlisted elements are zero. The elements of D_{in} and D_{out} are zero except for element $D_{in,1,1}$ which is ζ_3 and $D_{out,1,1}$ which is β_1.

Table 4.1 State, control, and output vector variable assignments

State vector		Control vector	Output vector
$x_1(\sigma_{12})$	$x_9(q_1)$	$u_1(U_{S1})$	$y_1(P_1)$
$x_2(\sigma_{23})$	$x_{10}(q_2)$	$u_2(U_{S2})$	$y_2(P_2)$
$x_3(S_1)$	$x_{11}(q_3)$	$u_3(U_{S3})$	$y_3(P_3)$
$x_4(S_2)$	$x_{12}(h_{in2})$	$u_4(U_{V1})$	$y_4(\sigma_{12})$
$x_5(S_3)$	$x_{13}(r_1)$	$u_5(U_{V2})$	$y_5(\sigma_{23})$
$x_6(V_1)$	$x_{14}(r_2)$	$u_6(U_{V3})$	$y_6(h_{out,1})$
$x_7(V_2)$	$x_{15}(r_3)$		$y_7(h_{out,2})$
$x_8(V_3)$	$x_{16}(h_{in3})$		$y_8(h_{out,3})$

Table 4.2 Elements of matrix A

$A_{1,1} = \zeta_1$	$A_{2,8} = \zeta_{16}$	$A_{9,8} = 0.25/\tau_{12}$
$A_{1,2} = \zeta_2$	$A_{2,12} = \zeta_{11}$	$A_{10,9} = 0.25/\tau_{12}$
$A_{1,3} = \zeta_5$	$A_{2,16} = \zeta_{12}$	$A_{11,10} = 0.25/\tau_{12}$
$A_{1,4} = \zeta_6$	$A_{3,3} = -1/\tau_S$	$A_{12,11} = 0.25/\tau_{12}$
$A_{1,6} = \zeta_7$	$A_{4,4} = -1/\tau_S$	$A_{13,13} = -0.25/\tau_{23}$
$A_{1,7} = \zeta_8$	$A_{5,5} = -1/\tau_S$	$A_{14,14} = -0.25/\tau_{23}$
$A_{1,12} = \zeta_4$	$A_{6,6} = -1/\tau_V$	$A_{15,15} = -0.25/\tau_{23}$
$A_{2,1} = \zeta_9$	$A_{7,7} = -1/\tau_V$	$A_{16,16} = -0.25/\tau_{23}$
$A_{2,2} = \zeta_{10}$	$A_{8,8} = -1/\tau_V$	$A_{13,12} = 0.25/\tau_{23}$
$A_{2,4} = \zeta_{13}$	$A_{9,9} = -0.25/\tau_{12}$	$A_{14,13} = 0.25/\tau_{23}$
$A_{2,5} = \zeta_{14}$	$A_{10,10} = -0.25/\tau_{12}$	$A_{15,14} = 0.25/\tau_{23}$
$A_{2,7} = \zeta_{15}$	$A_{11,11} = -0.25/\tau_{12}$	$A_{16,15} = 0.25/\tau_{23}$
	$A_{12,12} = -0.25/\tau_{12}$	

Table 4.3 Elements of matrix B

$B_{3,1} = 1/\tau_S$	$B_{6,4} = 1/\tau_V$
$B_{4,2} = 1/\tau_S$	$B_{7,5} = 1/\tau_V$
$B_{5,3} = 1/\tau_S$	$B_{8,6} = 1/\tau_V$

Table 4.4 Elements of matrix C		
	$C_{1,1} = \beta_3$	$C_{5,2} = 1$
	$C_{1,3} = \beta_2$	$C_{6,1} = \beta_{13}$
	$C_{2,1} = \beta_6$	$C_{6,3} = \beta_{12}$
	$C_{2,2} = \beta_7$	$C_{7,1} = \beta_{16}$
	$C_{2,4} = \beta_5$	$C_{7,2} = \beta_{17}$
	$C_{2,12} = \beta_4$	$C_{7,4} = \beta_{15}$
	$C_{3,2} = \beta_{10}$	$C_{7,12} = \beta_{14}$
	$C_{3,5} = \beta_9$	$C_{8,2} = \beta_{20}$
	$C_{3,16} = \beta_8$	$C_{8,5} = \beta_{19}$
	$C_{4,1} = 1$	$C_{8,16} = \beta_{18}$

4.3 The H^∞ Loop Shaping Approach

The H^∞ loop shaping approach has received attention as a method which could have the potential for improvement in the control of the tandem cold metal rolling process due to its capability for assuring performance in the face of the various uncertainties and disturbances which are present in this system, and which are a major issue in improving its performance. In this technique the traditional intuitive methods of classical control are used in conjunction with H^∞ optimization methods to produce a controller that assures stability and performance despite differences between the process model and the actual process, assuming that frequency-dependent bounds on these differences can be adequately described in the frequency domain by appropriate bounding functions. In this method the control designer describes in the frequency domain the desired performance by appending weighting functions to the plant model to generate a shaped control loop. H^∞ optimization methods then are applied to realize a robust controller. The method often is used for robust control of linear time-invariant MIMO systems. More detail regarding the basics of the H^∞ loop shaping approach is available in [6].

At first glance it would appear that this method is ideally suited for control of the tandem cold rolling process as it involves a systematic method that assures performance and reduces the effects of uncertainties and disturbances. Therefore in this section the application of this approach to the control of the tandem cold mill will be addressed, with an evaluation of its potential for improvement in performance in a practical setting. An objective is to present material that is oriented toward the application of the H^∞ loop shaping technique to tandem cold rolling in a practical sense so that the reader with a limited background in advanced control theory can obtain a reasonably good feel for the application. To this end much of the more rigorous supporting mathematical theory is omitted from the text but remains available in the cited references.

4.3.1 Traditional Loop Shaping Technique

The conventional idea of loop shaping as a design method for control of SISO systems has been around for many years. As early as 1945 classical loop shaping is described in [7] by Bode and later considered by Horowitz [8] as a tool for the synthesis of feedback control systems. In general the concepts of this simpler traditional method have some parallels in certain respects to the more complex H∞ method so that an understanding of the traditional method is expected to aid in the understanding of the H∞ loop shaping method. A brief review of the conventional method therefore is presented in what follows. The description of this method is based on a SISO system as shown in Figure 4.3, where the plant and controller transfer functions are taken to be linear and time-invariant and are represented respectively as G and K, where the controller K is to be obtained by design, and where r represents a reference signal, e represents an error signal, u is a signal to the plant actuator, y is the output, and d is a disturbance such as an unmodeled perturbation, a high frequency resonant condition, or noise.

Three associated transfer functions for this system are as described in (4.47)–(4.49) respectively. These are the loop transfer function L, the transmission transfer function T from r to y, and the sensitivity transfer function S from r to e,

$$L = GK, \tag{4.47}$$

$$T = \frac{GK}{1 + GK}, \tag{4.48}$$

$$S = \frac{1}{1 + GK}. \tag{4.49}$$

The idea of conventional loop shaping is to establish specifications that keep the magnitude and phase of the loop transfer function within certain bounds at every frequency. There are specifications that are established at three areas in the frequency domain as depicted in the example of Figure 4.4.

These three areas are the low frequency region, the crossover frequency region, and the high frequency region. In the low frequency region $|L(jw)| \geq l(w)$, where $l(\omega)$ represents a lower bound on the frequency response, it is desired to have good tracking in response to changes in r and low sensitivity to uncertainties in the plant so that the magnitude of L in this region is made large, the magnitude of S is small, and the magnitude of T is close to unity. At frequencies around the zero dB crossover frequency generally it is desired to have the slope of the response selected to provide stability and an acceptable transient response, which implies a good phase margin. In the high frequency region $|L(jw)| \leq u(w)$, where $u(\omega)$ represents an upper bound on the frequency response, it is desired to have the magnitude of L

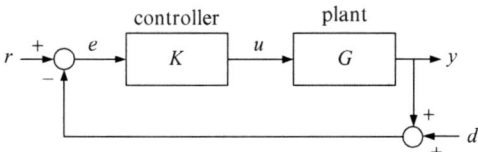

Fig. 4.3 SISO control system

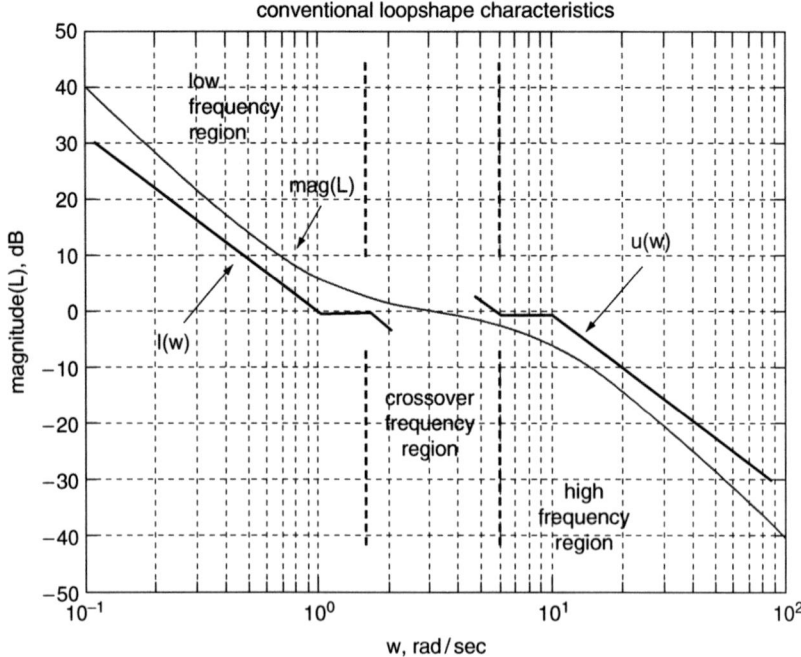

Fig. 4.4 Typical frequency domain areas, conventional loopshaping

be small so that T is small and the system is less sensitive to higher frequency disturbances such as noise or unmodeled perturbations, and that stability is retained if there are other unexpected characteristics such as small time delays or resonant points at these higher frequencies. Based on these requirements the frequency dependent bounds are established as shown in the example of Figure 4.4 and the controller K is designed by adjusting its dynamics until the desired frequency characteristic is obtained to be within the established bounds, or it is recognized that the loop shaping requirements may need to be adjusted or abandoned.

4.3.2 Matrix Singular Values

The understanding of the usage of the singular values of a matrix in their application to the H^∞ loop shaping approach to MIMO control is necessary to the comprehension of the H^∞ loop shaping method. Accordingly, a somewhat intuitive definition of the singular values is given in what follows, *i.e.*, for any complex $m \times p$ $(p \le m)$ matrix Q, the nonnegative square roots of the eigenvalues of Q^*Q are called the singular values of Q, where * indicates the complex conjugate transpose of Q. The set of singular values, the maximum (or upper) singular value, and the minimum (or lower) singular value are denoted as

$$\sigma(Q) = \{\sigma_i : i = 1, ..., p\}, \tag{4.50}$$

$$\bar{\sigma}(Q) = \sigma_1, \tag{4.51}$$

$$\underline{\sigma}(Q) = \sigma_p. \tag{4.52}$$

If the matrix Q is considered as a linear map from the vector space C^p to the vector space C^m which is defined by

$$\begin{aligned} Q &: C^p \mapsto C^m \\ Q &: u \mapsto Qu, \end{aligned} \tag{4.53}$$

then it can be shown that the upper and lower singular values are given as

$$\bar{\sigma}(Q) = \max_{\|u\|=1} \|Qu\|, \tag{4.54}$$

$$\underline{\sigma}(Q) = \min_{\|u\|=1} \|Qu\|, \tag{4.55}$$

where the norm is the vector Euclidean norm, so that $\bar{\sigma}(Q)$ and $\underline{\sigma}(Q)$ are the *maximum gain* and the *minimum gain* of the matrix Q. In a manner similar in certain respects to the traditional loop shaping method, in the H∞ loop shaping method the singular values are applied as a function of frequency to analyze various performance criteria and performance limitations. The use of singular values is considered further in the following sections.

4.3.3 H∞ Loop Shaping

As previously noted, the H∞ loop shaping technique can be used for development of controllers for linear time-invariant MIMO systems. Implementation using this technique includes the following three steps involving a plant G, controllers K_∞ and K, and performance weights W_1 and W_2, as depicted in Figure 4.5.

- *Step 1*: The desired shape of the plots of the frequency-dependent singular values of the control loop which includes the boundaries are determined using the intuition and experience of the designer. Weighting matrices then are added as a pre-compensator W_1 and a post-compensator W_2 to the plant G to come close to the desired shape of the open-loop singular values of the modified (*i.e.* shaped) plant.

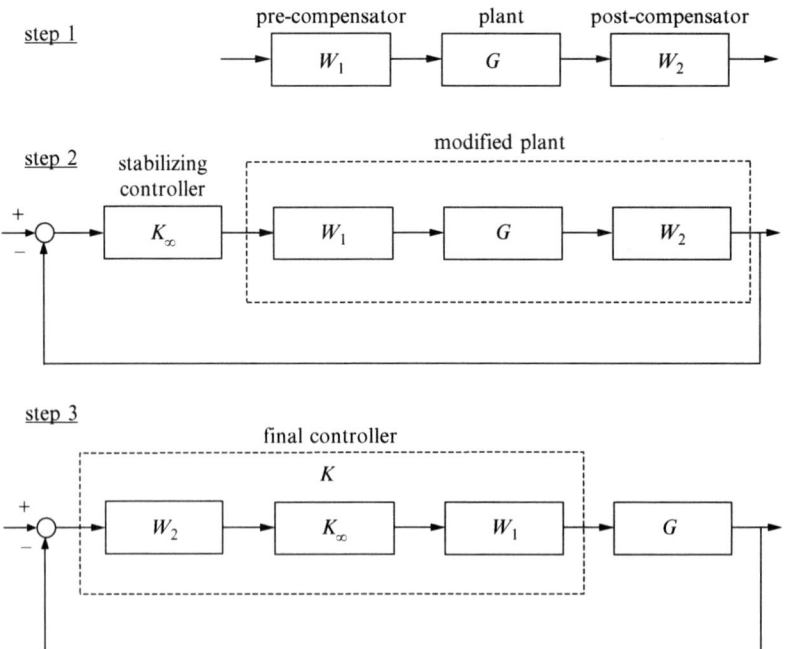

Fig. 4.5 Steps in H$^\infty$ loop shaping

- *Step 2*: H$^\infty$ optimization methods are applied to develop a controller K_∞ which stabilizes the shaped plant.
- *Step 3*: The final controller K for the original plant G is constructed by using the weights W_1 and W_2 and the stabilizing controller K_∞.

As noted in step 1, the desired loop shape and the boundaries are essentially established by intuition and the experience of the control designer, which often requires translating time response requirements of a MIMO system into the frequency domain, considering that frequency responses must be represented as plots of singular values. This can be a more than simple task, especially if the system is large and there are many possible combinations of uncertainties and disturbances that must be considered to form the boundaries on the frequency-dependent singular values.

The selection of the weights W_1 and W_2 is an even more daunting and difficult task, as a control practitioner has to develop a good deal of skill based on experience to be effective and obtain useful results. A set of guidelines for the selection of weights which was developed about 1991 by Hyde [9] for use in the design of control for aircraft has provided some help in this area. Since then some improvements in the guidelines have been made, mostly resulting from work in the aerospace industry. An example of recent work dealing with H$^\infty$ loop shaping and which offers some simplification in the procedure for the selection of weights is given in [10].

Step 2 is less difficult than step 1 as it involves determining a stabilizing controller K_∞ which satisfies (4.56),

$$\left\| \begin{bmatrix} I \\ K_\infty \end{bmatrix} [I - G_S K_\infty]^{-1} [I \quad G_S] \right\|_\infty < \gamma, \tag{4.56}$$

where G_S is the shaped plant, *i.e.* $G_S = W_2 G W_1$ with coefficients A, B, C, and D of its state-space realization, and for some $\gamma < \gamma_{min}$, with

$$\gamma_{min} = (1 + \lambda_{max}(XZ))^{1/2}. \tag{4.57}$$

where λ_{max} is the maximum eigenvalue of (XZ), and $X > 0, Z > 0$ are the unique solutions of the algebraic Riccati equations (4.58) and (4.59),

$$(A - BS^{-1}D'C)Z + Z(A - BS^{-1}D'C)' - ZC'R^{-1}CZ + BS^{-1}B' = 0, \tag{4.58}$$

$$(A - BS^{-1}D'C)'X + X(A - BS^{-1}D'C) - XBS^{-1}B'X + C'R^{-1}C = 0, \tag{4.59}$$

and where

$$R = I + DD', \; S = I + D'D \tag{4.60}$$

The state-space realization of the controller K_∞ is given by (4.61)–(4.66) as

$$A_{K\infty} = A + BF + \gamma^2(L')^{-1}ZC'(C + DF), \tag{4.61}$$

$$B_{K\infty} = \gamma^2(L')^{-1}ZC', \tag{4.62}$$

$$C_{K\infty} = B'X, \tag{4.63}$$

$$D_{K\infty} = -D', \tag{4.64}$$

where

$$F = -S'(D'C + B'X), \tag{4.65}$$

and

$$L = (1 - \gamma^2)I + XZ. \tag{4.66}$$

The final controller K is then determined as in step 3,

$$K = W_1 K_\infty W_2. \tag{4.67}$$

It should be noted that the computations in step 2 above can be done by using the software tools provided in software packages such as MATLAB/Simulink.[2] Also, as noted in [10] based on general practice in the aerospace industry, if $\gamma_{min} < 4$ the design generally will be successful, but if $\gamma_{min} > 4$ then the augmented plant is incompatible with closed-loop system robust stability and the weighting functions most likely will require modification.

4.3.4   Application to Tandem Cold Metal Rolling

The H^∞ loop shaping technique has been investigated [11, 12] for its application to the control of centerline thickness and interstand tension for a tandem cold metal rolling application. As part of this investigation, a controller for a typical tandem rolling mill was designed and its performance was simulated and compared with a typical conventional controller, similar to that shown in Figure 3.10. The design of the controller generally followed the pattern outlined previously, *i.e.* a nonlinear model and a linearized version of it were developed and verified, system inputs and outputs were selected, a plant structure was established, the boundaries within which the plots of the open-loop singular values should remain were established to assure system performance and robustness, loop shaping weights were selected, simulation was performed and the results evaluated. To give some insight into the issues that might be encountered by a control practitioner in an actual design process using the H^∞ loop shaping technique, some of the more significant aspects of the investigation into the use of this method for control of the tandem cold mill are described in what follows.

The selection of inputs is fixed by the available actuators so that any variation in this selection process is in the outputs. Several criteria were considered in the selection of the outputs. In accordance with guidance of [9] it was desired to attempt to keep the plant as diagonal as possible so as to keep the loop shaping weights as diagonal as possible to simplify the system design. This would be especially important in a larger plant such as the tandem cold rolling mill, as the use of off-diagonal weights would add considerably to the number of tuning parameters. The association of outputs to inputs was determined by the use of the relative gain array method [13] by which the interactions between process variables can be assessed. In this method a matrix is formed wherein the impact of a control variable on the output variables, relative to its impact on the other variables, for all possible input-to-output pairings is determined. The idea is to select the pairings of the variables so as to match an input with a certain output that is most affected by a change in the input, while reducing any undesirable effects on the other variables. Table 4.5 lists the resulting input–output pairings.

---

[2]MATLAB and Simulink are registered trademarks of The MathWorks, Inc., Natick, MA 01760-2098.

Table 4.5 Pairing of process inputs and outputs, based on [11]

Inputs	Outputs
Stand 1 roll gap actuator	Stand 1 thickness gauge
Stand 2 roll gap actuator	Stand 2 estimated thickness
Stand 3 roll gap actuator	Stands 2,3 tension
Stand 4 roll gap actuator	Stands 3,4 tension
Stand 5 roll gap actuator	Stand 5 roll force
Stand 1 roll speed actuator	Stand 1 peripheral speed
Stand 2 roll speed actuator	Stands 1,2 tension
Stand 3 roll speed actuator	Stand 3 estimated thickness
Stand 4 roll speed actuator	Stands 4,5 tension
Stand 5 roll speed actuator	Stand 5 thickness gauge

Due to the size of the system, the total number of states was 57, prior to attempts to reduce the order of the model. After applying model order reduction techniques, the number of states was reduced to about half that number.

The selection of input and output weights for loop shaping is a major task in the design effort. In the design that was performed as described in [11, 12] the guidelines developed in [9] were used as a general basis for weight selection, although in more recent loop shaping applications improvements developed in the guidelines would make the difficult task of weight selection a bit easier. The specific guidelines for weight selection that were used in [11] are:

- *Guideline 1*: Scale all outputs such that one unit of cross coupling into each of the outputs is equally undesireable.
- *Guideline 2*: Scale all inputs to reflect the relative actuator bandwidth capabilities, which depend on the dynamics of the actuator bandwidth and the saturation limits.
- *Guideline 3*: Arrange the inputs and outputs so that the plant is as diagonal as possible, which simplifies the final design as the loop shaping can be done using diagonal weights.
- *Guideline 4*: Adjust the roll-off rates of the singular values at the cross-over frequency by plotting the transfer functions from each actuator alone to all of the outputs, and then from all of the actuators to each output alone. A desired roll-off rate at cross-over is about −20dB/decade at the desired bandwidth. Integral action at the precompensator and roll-off terms at higher frequencies can be added as needed.
- *Guideline 5*: Align the singular values at the desired bandwidth, assuming the plant is well-conditioned.

The selection of the output weighting (*i.e.* weight W_2) requires intuition into the characteristics of the physical plant and familiarity with its operation. In the case of the application of guideline 1, one unit of cross coupling was taken to be the acceptable tolerance in each output variable as represented in the scaled system. For example for the exit thickness in this case a tolerance of 1% was equivalent to a 0.02 mm variation in the actual thickness so that the preliminary scaling for this output was 1/0.02. Table 4.6 lists the acceptable tolerances used for other variables.

Table 4.6 Tolerance used in scaling of output weights, based on [11]

Variable	Tolerance (%)
Measured thicknesses	1
Interstand tension forces	10
Roll forces	10
Work roll speeds	1

It should be noted that these particular tolerances were used in this specific instance and that in other applications different, and possibly tighter, tolerances may be more consistent with the actual equipment and system characteristics.

Input weights also require scaling. In the case of the roll gap actuators, the inputs were scaled such that the smallest step change in the actuator movement which produced unacceptable transients in the open-loop model of the mill was taken to represent one unit of scaling. For the gap actuators this was 0.1 mm which was scaled as 0.1. In the case of the work roll drives, scaling was such that 1% of the nominal drive speed represents an acceptable activity which in this case was a deviation of about 1% in the exit thickness. For this application all drives were scaled identically. The scaling of the weightings related to the estimates of the stand output thicknesses using the BISRA gaugemeter method (Section 3.4.3.4) was made deliberately lower as it was considered that the estimates are less reliable. In some cases it was noted that iteration was required to tune the weight scaling during simulation. Each of the elements of weight W_1 was a simple low pass filter, and in accordance with Guideline 4 integral action was added to each of these elements which improved the steady-state accuracy.

Considerable tuning was required to come up with the final configuration of the weights. Some of the issues encountered were excessive movements in the roll gap actuators that produced thickness variations that were difficult to correct downstream. Also, the coupling of roll eccentricity into the responses of the downstream drives near exit of the mill was seen in the speed of the exit strip which could affect the exit coiler controllers. These issues were corrected somewhat by the adjustment of the integral gain in the roll actuator weights and the addition of high frequency roll-off in the drive actuators weights. In [11] several plots of the singular values after final tuning of the weights were presented and an acceptable final value for γ_{min} was realized.

During simulation the controller was coupled to a nonlinear model of the plant to verify final performance. It was noted that tuning of the loop shaping controllers during simulation to achieve acceptable results was less than straightforward, and there was difficulty in determining how the adjustment of the weighting elements would affect the transient responses which required significant redesign to correct so that an improvement in performance over the conventional control could be attained. This was attributed mostly to the nonlinear characteristics of the mill as represented in the nonlinear model.

The simulation results for responses to typical disturbances in the incoming thickness and hardness were compared to those generated using a conventional controller. At top speed the mill exit thickness was held to within −1% to +1% of nominal thickness while the conventional controller as simulated for this investigation held thickness variations to within -2.5% to $+0.5\%$ of nominal thickness, which shows an improvement. Interstand tension control was deemed adequate. During speed change from thread to run and the reverse a series of controllers were designed and gain scheduled to control the mill with a resulting slight degradation in performance, so that the tolerance in exit thickness was held to within -1.5% to $+1.5\%$ of nominal thickness during these regimes of operation. Several plots of the results of the various simulations are available in [11, 12].

4.3.5 Comments

The preceding has been included to provide an example of an advanced technique as applied to the tandem cold metal rolling process. As with the application of almost any new method, there are areas that show considerable advancement while there are others that possibly could be improved by further research and development efforts. While considerable effort as described in [11] was required to generate and simulate the H$^\infty$ loop shaping controller, on an overall basis it is considered that somewhat more work is needed for this technique to be judged as acceptable for usage in a practical setting. Among the main issues is the multitude of different products that are processed during a typical working period. The controller design considered was developed around only one product. To be useful the controller must be capable of handling a wide range of products wherein a change can occur rapidly (and often unexpectedly). This is especially important in the case of continuous tandem cold rolling as the strip characteristics can change in milliseconds during the transition from the tail of one coil to the head of the next coil, and considering that the present trend in new rolling applications and in the revamp of existing mills is toward continuous tandem cold rolling as it produces significantly higher yield and profitability.

Another area of concern is user friendliness to design and commissioning personnel who have a limited background in advanced control theory, especially in the theory and application of H$^\infty$ robust control techniques. This is especially significant as it is crucial that any control strategy applied to this process must be useful in a practical sense as user friendliness plus simplicity and ease of tuning at commissioning are important to a rapid startup and also later during actual operation, which ultimately affects production and profitability. The dynamic MIMO nature of the controller is somewhat less than desirable for commissioning since personnel involved in this area generally prefer a simple control structure wherein a tuning adjustment is related to a single tunable variable that is easily related to a single process function. Also, during the simulations it was noted that tuning of the weights was difficult even for those with considerable talent and background in the development of these types of controllers so that tuning by those with only a basic background in control theory could be a bit more daunting.

Moreover, the controller requires a linear model which adds to the design complexity, especially to accommodate a change in mill speed from thread to run and the reverse.

Thus there is considerable room for improvement in these areas. However, future developments may result in further advances that prove to be beneficial in the application of H^∞ loop shaping techniques to this important industrial process, so that the potential of this method to improve product quality as well as being useful in a practical sense might be realized in future investigations.

4.4   An Observer-based Method for MIMO Control

In this section a MIMO design for control of the tandem cold rolling process using an observer-based technique is presented. The process is a two-stand tandem cold mill. A design based on the use of this method for control of this process has been simulated and successfully applied to the control of an actual mill. The significance of considering this observer-based technique for a MIMO controller is its capability to successfully handle disturbances in the presence of uncertainties, which is highly important in the control of this process. The basic concepts for observers in MIMO systems, the internal model principle, and feedforward techniques tie in with the overall approach to the controller design and therefore a brief overview of each of these is included as useful background which can aid in the understanding of the controller design. What is included is intended to be oriented toward practical application and therefore much of the mathematical supporting theory is omitted but can be accessed through the references cited.

4.4.1   Observers for Linear Systems

There are many types of observers that are used to estimate the states and disturbances of linear systems. As an introduction to the consideration of the controller design for the tandem cold mill and as an aid to the reader, some supporting material related to observers is presented.

The Luenberger identity observer [14] is very much like the observer utilized in the design of the controller for the tandem mill, and therefore it will be considered. The intent of an observer, whether an identity observer or one of another type, is to produce an approximation to the full state vector using the control variables and the output variables of the original system. The observer in itself is a linear dynamical system. The intent of the state vector of the observer is to have its states generate the information that is missing about the states of the original system. The observer is in fact a dynamic entity that generates the full state vector.

To better understand the identity observer, the following system is assumed to be completely observable (the definition of observability can be found in standard texts on linear system theory, *e.g.* [5]),

$$\dot{x} = Ax + Bu, \ x(0) = x_0, \tag{4.68}$$

$$y = Cx, \tag{4.69}$$

where $x \in R^n$ is the state vector, $y \in R^p$ is the output vector, $u \in R^m$ is the control vector, and A, B, and C are constant matrices of appropriate dimensions. An observer for this system can be constructed which is of the form

$$\dot{z} = Az + E(y - Cz) + Bu, \ z(0) = z_0, \tag{4.70}$$

where $z \in R^n$ is the state vector of the observer and $E \in R^{n \times p}$ is a constant matrix. Figure 4.6 is a schematic of the identity observer, with inputs u and y from the original system.

Equations 4.69 and 4.70 can be combined to get (4.71) and (4.72) which generate an observer error vector e,

$$\dot{e} = (A - EC)e, \ e(0) = e_0, \tag{4.71}$$

where

$$e = z - x. \tag{4.72}$$

This type of observer is denoted as an identity observer since its state z tracks the state x of the original system. If at $t = 0$ the error e is zero, then for $t > 0$ the error is zero. If the initial error is not zero then the error for $t > 0$ is determined as in (4.71). If the matrix $(A–EC)$ is asymptotically stable, the error will tend toward zero at a rate that is determined by the dominant eigenvalue of $(A–EC)$. In the design of the

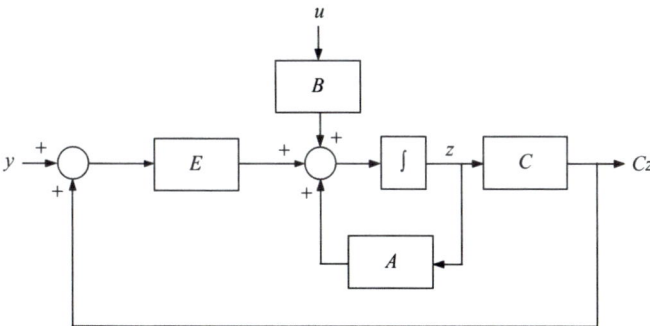

Fig. 4.6 Schematic of an identity observer

controller for the tandem mill, an identity observer is utilized to estimate the
disturbances and the states of the system.

4.4.2   *The Internal Model Principle*

The internal model principle has been in use for a long time in the synthesis of
controllers as an effective means for mitigating the effects of disturbances, and at
the same time tracking a desired reference in the presence of various uncertainties.
The initial consideration of the application of the internal model principle to SISO
systems is presented to give some insight into its general use. Understanding the
application to SISO systems provides a basis for a better understanding of the ideas
used in certain MIMO systems.

As presented in [15] the internal model principle can be summarized by stating
that a controller synthesis is structurally stable (as defined in [15]) only if the
controller utilizes feedback of the controlled variable, and incorporates in the
feedback path a correctly duplicated model of the dynamic structure of the external
signals which the controller is required to process. How this applies for the internal
compensation of an external disturbance in the case of steady-state can be seen in
what follows using Figure 4.7 which depicts a SISO control loop with a disturbance.
In Figure 4.7 Laplace transforms are shown for the reference signal as $R(s)$, the
control signal as $U(s)$, the output as $Y(s)$, the disturbance as $D(s)$, with $C(s)$ as the
transform of the controller, and $G(s)$ of the plant.

Assuming that the numerator and denominator of $C(s)$, $G(s)$, and $D(s)$ can be
factored as

$$C(s) = \frac{C_n(s)}{C_d(s)}, \tag{4.73}$$

$$G(s) = \frac{G_n(s)}{G_d(s)}, \tag{4.74}$$

$$D(s) = \frac{D_n(s)}{D_d(s)}, \tag{4.75}$$

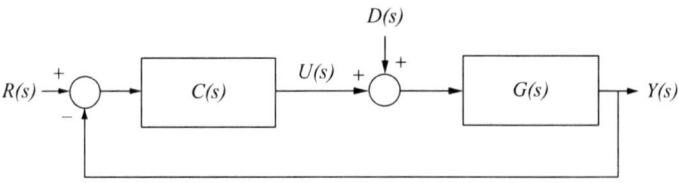

Fig. 4.7 SISO control loop with a disturbance

where $C_n(s)$ and $C_d(s)$ are the factors of the numerator and denominator of $C(s)$, and similarly for $G(s)$ and $D(s)$, and if the factors of $D_d(s)$ are also the factors of $C_d(s)$, then in the general case the steady-state error in the output $Y(s)$ due to the disturbance is zero. This can be seen in (4.76)–(4.78) where

$$Y(s) = \frac{G(s)}{1 + G(s)C(s)} D(s), \tag{4.76}$$

$$Y(s) = \frac{G_n(s)}{G_d(s)C_d(s) + G_n(s)C_n(s)} D_n(s), \tag{4.77}$$

and the steady-state value of $y(t)$ is

$$y(\infty) = \lim_{t \to \infty} y(t) = \lim_{s \to 0} \left(\frac{s\, G_n(s) D_n(s)}{G_d(s)C_d(s) + G_n(s)C_n(s)} \right) = 0. \tag{4.78}$$

It also can be noted that under the conditions of robust stability of the nominal control loop (*i.e.*, the control loop excluding uncertainties in the model), if the factors of $D_d(s)$ lie in the closed right-half plane, then these unstable components also will be compensated. Moreover, the output $U(s)$ of the controller generally will contain the modes of the disturbance.

In summary, as shown in the above SISO example, the general idea of the internal model principle is that if a controller is to be any good, it must include a model of the external world. This carries over in a general sense in what follows in Section 4.4.4 for the MIMO controller for the two-stand tandem mill, wherein the disturbances are modeled as the outputs of an observer that reflects the nature of the disturbances, and are used along with the estimated states to form the final control law for the system.

4.4.3 Feedforward Techniques

Feedforward techniques are often used for compensation of disturbances and this is the case for control of the two-stand tandem cold mill. An example of a feedforward method is depicted in the SISO system of Figure 4.8.

In Figure 4.8 the disturbance $D(s)$ is assumed to be measureable. The idea is to design controllers $C_1(s)$ and $C_2(s)$ so that $D(s)$ has no effect on the output $Y(s)$. It is assumed that the controller $C(s)$ is designed so that the system is stable and has an acceptable response assuming the disturbance $D(s)$ is zero. If $D(s)$ is nonzero, the output $Y(s)$ due to the effects of the disturbance is then

$$Y(s) = \frac{C_2(s) - C_1(s)G(s)}{1 + C(s)G(s)} D(s). \tag{4.79}$$

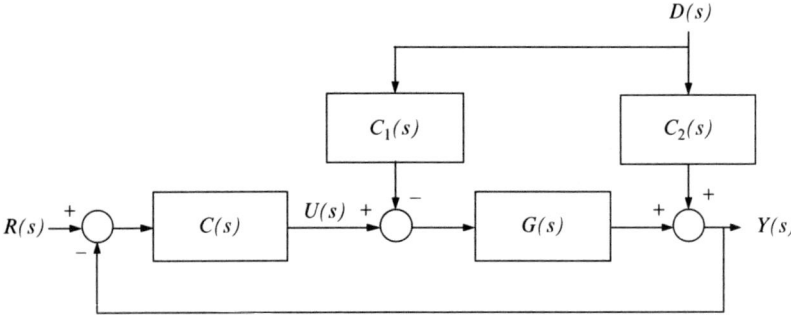

Fig. 4.8 Feedforward disturbance compensation

From (4.79) it can be seen that the relationship between controllers $C_1(s)$ and $C_2(s)$ takes the form of (4.80) to cancel the disturbance $D(s)$.

$$C_1(s) = \frac{C_2(s)}{G(s)}. \tag{4.80}$$

Somewhat similar techniques of disturbance rejection using feedforward methods for disturbance compensation can be extended to MIMO systems as will be seen in the presentation of the design of the control for the tandem cold mill in Section 4.4.4. It also can be noted that while the internal model principle provides compensation for disturbances in steady-state, the use of feedforward provides compensation during the transient portion of the response.

4.4.4 MIMO Controller for a Two-Stand Tandem Cold Mill

The two-stand tandem cold mill is designed to process aluminum which poses somewhat more stringent requirements on the control than in the case of steel. This is because the material is softer than steel and therefore is more susceptible to disturbances such as changes in the friction coefficient, uncertainties in the compressive yield stress (hardness), plus others. The controller for this process is a MIMO configuration as described in [16] and illustrates some of the concepts involved in MIMO control of tandem cold rolling. Figure 4.9 presents a schematic of the mill.

The objectives of the controller for this process are to maintain the centerline thickness at the mill exit within a desired tolerance and to reduce excursions in the interstand thickness and tension to support the stability of rolling. The nonlinear model used essentially follows that presented in Section 4.2 and Chapter 2, with appropriate adjustments for a two-stand rolling application. The basic equations of the nonlinear model, with some minor modifications for this application, are listed for the convenience of the reader. The variables represented by the symbols used

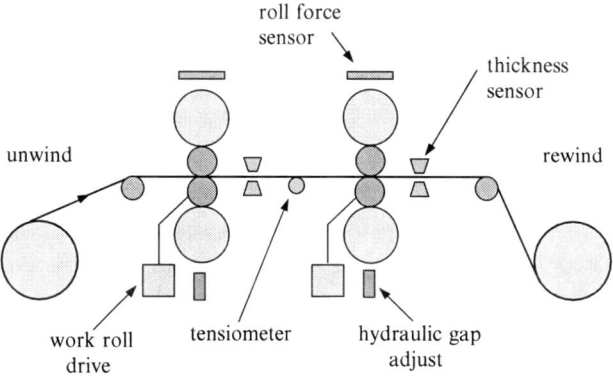

Fig. 4.9 Schematic of a 2-stand tandem cold mill, based on [16]

are as noted in Chapter 2. The relationships given in (4.81)–(4.89) describe respectively the strip output thickness, the interstand tension, the specific roll force, the strip speed at the stand output, the forward slip, the mass flow across the roll gap, the roll gap actuator position, the work roll actuator speed, and the stand input thickness.

$$h_{out} = S + S_0 + \frac{PW}{M}, \tag{4.81}$$

$$\frac{d(\sigma_{i,i+1})}{dt} \equiv \dot{\sigma}_{i,i+1} = \frac{E(V_{in,i+1} - V_{out,i})}{L_0}, \quad \sigma_{i,i+1}(0) = \sigma_{0,i,i+1}, \tag{4.82}$$

$$P_i = P_i(h_{in}, h_{out}, \sigma_{in}, \sigma_{out}, \bar{k}, \mu), \tag{4.83}$$

$$V_{out,i} = V(1+f), \tag{4.84}$$

$$f_i = f_i(h_{in}, h_{out}, \sigma_{in}, \sigma_{out}, \bar{k}, \mu), \tag{4.85}$$

$$MF_i = V_{in}h_{in} = V_{out}h_{out}, \tag{4.86}$$

$$\dot{S}_{p,i} = \frac{U_{S,i}}{\tau_S} - \frac{S_{p,i}}{\tau_S}, \quad S_{p,i}(0) = S_{p,i,0}, \tag{4.87}$$

$$\dot{V}_{p,i} = \frac{U_{V,i}}{\tau_V} - \frac{V_{p,i}}{\tau_V}, \quad V_{p,i}(0) = V_{p,i,0}, \tag{4.88}$$

$$h_{in,i+1}(t) = h_{out,i}(t - \tau_{d,i,i+1}). \tag{4.89}$$

Based on these nonlinear representations, a linearized process model as given in (4.90)–(4.92) was developed that includes disturbances and uncertainties.

$$\frac{dx}{dt} \equiv \dot{x} = Ax + Bu + D_1 d, \tag{4.90}$$

$$y_1 = C_1 x + D_2 d, \tag{4.91}$$

$$y_2 = C_2 x + D_3 d, \tag{4.92}$$

where $x \in R^4$, $y_1 \in R^3$, $y_2 \in R^2$, $u \in R^3$, are respectively vectors whose elements represent the state, output 1, output 2, and control variables, with $d \in R^5$ being a vector whose elements represent the disturbances and uncertainties, and with A, B, C_1, C_2, D_1, D_2, and D_3 being matrices of appropriate dimensions whose elements are constants as determined for a particular operating point as noted in [16]. In this case the particular uncertainties considered are represented as disturbances, and all variables in the linearized equations are understood to represent excursions from nominal operating point values. Further in (4.84) and (4.85) the roll gap positions $S_{p,i}$ and work roll peripheral speeds $V_{p,i}$ are the actuator outputs as modified by certain disturbances such as set-up errors and roll thermal expansions. Table 4.7 lists the process variables that are assigned to the elements of x, y, and u, where $F1 = P_1 W/M$ with P_1 being the specific roll force for stand 1 and W and M being the strip width and the mill modulus, and $F2$ being similarly understood. Table 4.8 lists the elements of the vector d that are associated with the individual disturbances. It should be noted that in Table 4.8 an individual disturbance may affect more than one element of d. In this model the stand 2 work roll peripheral speed is not identified as a variable assigned to an element of the state vector as it was considered that the controlled variables are not directly affected by changes in this variable. The simulation results are based on this assumption.

In this design the poles of the plant that correspond to the time constants of the actuator control loops are adjusted to assure a reasonably fast response of the actuators. This is done by state feedback as depicted for example in Figure 4.10 where state feedback is applied to move the pole related to the actuator time constant τ_{V1} for the roll peripheral speed to a new pole related to the shifted time constant $\hat{\tau}_{V1}$, where $\hat{U}_{V1}$ is an input which is used for disturbance compensation and V_{p1} is understood to be the stand 1 work roll peripheral speed.

The inputs $\hat{U}_{V,i}$ (for work roll peripheral speed) and $\hat{U}_{S,i}$ (for roll gap position) are feedforward control signals determined by functions denoted herein in Figure 4.11 as $f(d,p)$, where p represents parameters that relate the elements of the disturbance vector d to the controllers for the individual actuators. These functions are of the

Table 4.7 Variable assignments for state, control, and output vectors, based on [16]

State vector	Control vector	Output vector 1	Output vector 2
x_1 (σ_{12})	u_1 (U_{S1})	y_{11} (h_{out1})	y_{21} ($F1$)
x_2 (S_{p1})	u_2 (U_{V1})	y_{12} (σ_{12})	y_{22} ($F2$)
x_3 (V_{p1})	u_3 (U_{S2})	y_{13} (h_{out2})	
x_4 (S_{p2})			

Table 4.8 Elements of the d vector associated with each disturbance, based on [16]

Disturbance	Associated elements of the d vector
Stand 1 input thickness	d_1, d_2
Stand 2 input thickness	d_1, d_3
Mill input hardness	d_1, d_2, d_3
Stand 1 friction	d_1, d_2
Stand 2 friction	d_1, d_3
Stand 1 roll eccentricity	d_4
Stand 2 roll eccentricity	d_5
Mill input tension	d_1, d_2
Mill output tension	d_1, d_3
Stand 1 roll wear	d_4
Stand 2 roll wear	d_5
Stand 1 roll thermal expansion	d_4
Stand 2 roll thermal expansion	d_5
Stand 1 roll gap set-up	d_4
Stand 2 roll gap set-up	d_5

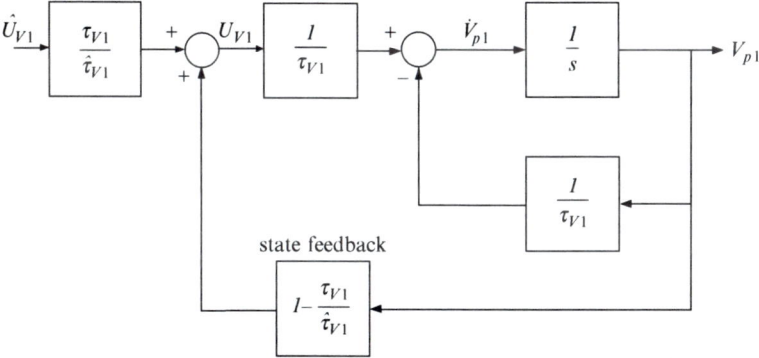

Fig. 4.10 An example of the application of state feedback for work roll peripheral speed

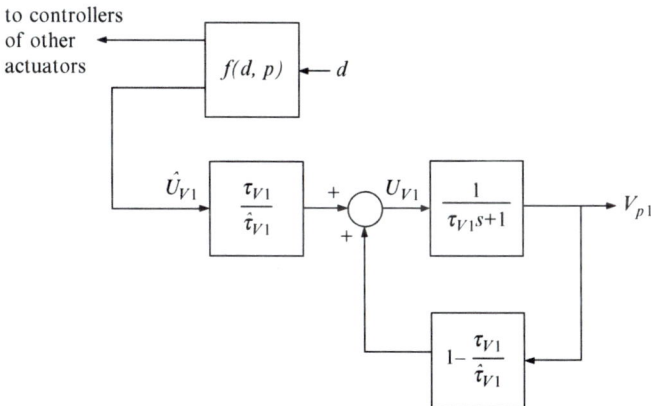

Fig. 4.11 Example of disturbance mitigation using work roll peripheral speed for feedforward compensation

type that are derived in [16] for compensation of disturbances based on the effects of the various actuators on each of the disturbances as noted in Table 4.9, and with the requirement that the excursions in the exit thicknesses and the interstand tensions are to be zero. In certain instances as noted in Table 4.9 compensation for a disturbance requires the concurrent action of more than one controller. An example of the use of work roll peripheral speed for feedforward compensation for mitigation of disturbances is depicted in Figure 4.11.

Since the individual states and the disturbances are not measurable, they are estimated using the identity observer. The final form of the controller is then determined using the estimated states and the estimated disturbances. How the identity observer can be used to estimate elements d_4 and x_2 of the disturbance and state vectors is given in (4.93),

$$\frac{d}{dt}\begin{bmatrix} z_1 \\ z_2 \end{bmatrix} = a_{S1}\begin{bmatrix} z_1 \\ z_2 \end{bmatrix} + b_{S1}U_{S1} + \begin{bmatrix} k_1 \\ k_2 \end{bmatrix}(y_{11} - y_{11,e}), \tag{4.93}$$

where z_1 and z_2 are elements of the vector $z \in R^6$ of the observer, with z_1 being the estimate of d_4 and z_2 being the estimate of x_2 (Table 4.7), k_1 and k_2 are elements of the observer gain vector $k \in R^6$, y_{11} is an element of the vector $y_1 \in R^3$, and $y_{11,e}$ is an estimate of y_{11}, with a_{S1} and b_{S1} being constants of appropriate dimensions. Similar expressions are derived which involve the remainder of the state and disturbance vectors. With some assumptions on the disturbances and algebraic manipulations using derived relationships, the resulting form of the final description [16] of the process controller is

$$\dot{z} = Fz + Gy_1 + Hy_2, \tag{4.94}$$

$$u = Pz + Qy_1 + Ry_2, \tag{4.95}$$

where the elements of z are as given in Table 4.10, the elements of vectors u, y_1, and y_2 are as shown in Table 4.7, and F, G, H, P, Q, and R are constant matrices.

In Table 4.10 τ_σ is the time constant associated with the linearized model of the tension, while disturbances d_1 and d_2 can be obtained from disturbances d_4 and d_5 and therefore have been incorporated in the controller through the estimates of d_4 and d_5.

System stability was analyzed as part of the design described in [16]. The criteria applied for stability for this MIMO system utilizes the function described by the characteristic equation, *i.e.* if this function encircles the origin the system is closed-loop

Table 4.9 Control variables for compensation of disturbances, based on [16]	Disturbance	Control variable for disturbance compensation
	d_1	U_{V1}
	d_2	U_{V1}, U_{S1} (concurrently)
	d_3	U_{V1}, U_{S2} (concurrently)
	d_4	U_{S2}
	d_5	U_{S1}

Table 4.10 Estimated variables in the final form of the controller	Element of z	Estimated variable
	z_1	d_4
	z_2	x_2
	z_3	$x_3 + k_3\tau_\sigma x_1$
	z_4	$d_1 + k_4\tau_\sigma x_1$
	z_5	d_5
	z_6	x_4

unstable. Characteristic equations for the major loop and the minor loop of the system were considered for positive frequencies between zero and infinity. In both cases the major and minor loops were stable. Additional criteria regarding the use of the characteristic equation in determining stability of linear MIMO systems is found in advanced texts on linear MIMO system control [e.g. 17].

Simulations were performed with the MIMO system coupled to the model, and compared with conventional controllers using SISO control loops. The comparisons were based on disturbances which included excursions in the friction coefficients based on the changes in the speed of the work rolls, on eccentricity in the backup rolls, and on performance during speed change from thread speed to run speed. The major improvement over the conventional method was the reductions in the excursions in the exit thickness of both stand 1 and stand 2 during acceleration. In the case of stand 1, the deviation in the exit thickness was about 15 μm with the conventional control and about 5 μm with the MIMO control. Similar results were noted for stand 2 where the deviation was about 30 μm with the conventional method and 5 μm with the new method. During acceleration and at steady speed the excursions in tension were about the same in both cases, and at steady speed operation both controllers held the exit thicknesses to within 3 μm at the exit of stand 1 and 2 μm at the exit of stand 2, with the performance of the MIMO controller being slightly better than that of the conventional controller.

Additional simulation was performed to compare the frequency responses to each of the disturbances for the new controller to the conventional controller, where for each disturbance the resulting excursions in thicknesses and tension were determined. In almost every case examined there was a reasonable improvement, especially at the lower frequencies.

The new controller was installed on an actual mill where data during typical operational modes were obtained. The data were mostly consistent with the improvements noted during the simulation. During the acceleration and deceleration phases for a 0.32 mm nominal mill exit thickness the strip length that was outside the desired tolerance of +/−0.004 mm was reduced by about 60%, which is a significant improvement.

4.4.5 Comments

In general MIMO controllers are able to deal inherently with interactions between the various process variables which is an advantage over a controller consisting only of several SISO control loops. This generally is the case for the MIMO design just presented for the 2-stand tandem mill. As shown in the simulation results presented in [16] and in the application to an actual operating system, there were improvements in performance over a conventional method which offer a confirmation that a MIMO based controller provides an improvement over a SISO based controller for this application. Moreover, these results imply a strong possibility for improving performance when a MIMO system is extended to the control of a mill with more than two-stands. It is considered that some additional advantages could be realized by treating the interstand time delays as part of the actual model rather than as disturbances which would lead to a more realistic simulation and possible further improvement in performance. Other considerations would involve designing the controller to perform well without a thickness sensor located between adjacent mill stands, as this is typical for most mills with additional stands, expanding the design to treat the uncertainties separately from the disturbances and including a more complete range of uncertainties in modeling and measurement, and fully considering in the model the work roll peripheral speed of the last stand. Additionally, the design is based on a linear system which adds a bit of complexity for linearization and may require some additional gain scheduling methods to accommodate changes in speed or product, although it is not mentioned in [16] if such methods are incorporated into the design for the speed changes that were described. As in the case of other MIMO based systems, the tuning of the more complex MIMO controllers at commissioning can present difficulties especially for personnel with only a basic background in control theory. Nevertheless, the consideration of the MIMO design has shown to be an improvement over the SISO method, and with some additional development can offer an even better approach to control of the tandem mill.

4.5 Concluding Comments

This chapter has presented two advanced techniques for control of the tandem cold rolling process. The advantages and disadvantages of each were discussed. While each method offered certain advantages over a conventional controller, nevertheless there were some also disadvantages that made the advanced control method somewhat less desirable. Chapter 5 presents a third method that offers significant improvement over conventional control without most of the disadvantages of the two methods presented.

References

1. Kuo BC. Automatic control systems. Englewood Cliffs: Prentice-Hall; 1991.
2. Buckley PS. Techniques of process control. New York: Wiley; 1964.
3. Vajta M. Some remarks on Pade approximations. Proceedings of the 3rd TEMPUS-INTCOM Symposium; 2000; Veszprem.

4. Yoon MG, Lee BH. A new approximation method for time delay systems. IEEE Trans Automat Control. 1997;42(7):1008–12.
5. DeCarlo RA. Linear systems, state-space methods. Englewood Cliffs: Prentice-Hall; 1989.
6. McFarlane D, Glover K. A loop shaping design procedure using H^∞ synthesis. IEEE Trans Automat Control. 1992;37(6):759–69.
7. Bode HW. Network analysis and feedback amplifier design. New York: Van Nostrand; 1945.
8. Horowitz JM. Synthesis of feedback systems. New York: Academic Press; 1963.
9. Hyde RA, Glover K. The application of scheduled H^∞ controllers to a VSTOL aircraft. IEEE Trans Automat Control. 1993;38(7):1021–39.
10. Panesi R. H^∞ loop shaping control of aerospace systems, PhD thesis. Pisa: University di Pisa; 2008.
11. Geddes EJM. Tandem cold rolling and robust multivariable control, PhD thesis. Leicester: University of Leicester; 1998.
12. Geddes EJM, Postlewaite I. Improvements in product quality in tandem cold rolling using robust multivariable control. IEEE Trans Control Syst. Technol. 1998;6(2):257–69.
13. Maciejowski JM. Multivariable feedback design. Wokingham: Addison-Wesley; 1989.
14. Luenberger DG. Introduction to dynamic systems. New York: Wiley; 1979.
15. Francis BA, Wonham W. The internal model principle of control theory. Automatica. 1976; 12:457–65.
16. Hoshino I, Maekawa Y, Fujimoto T, et al. Observer-based multivariable control of the aluminum cold tandem mill. Automatica. 1988;24(6):741–54.
17. Corriou JP. Process control, theory and applications. London: Springer; 2004.

Chapter 5
Advanced Control: The Augmented SDRE Technique

5.1 Background

The advanced method presented in this chapter is the result of several years work at the University of Pittsburgh related to discovering an improved method for the control of the centerline thicknesses and the tensions in the tandem cold metal rolling process, with the aim of improving the quality and yield of the final product. As a major part of this work, the state-dependent Riccati equation (SDRE) technique was investigated as to having a significant potential for realizing this improvement. The results of this investigation were highly successful as will be seen in the following portions of this chapter wherein the results of simulations for both stand-alone and continuous tandem cold rolling mills are presented and compared with available data from actual installations, most of which are controlled by systems that are essentially SISO. The existing systems have done a reasonably good job in controlling the process, producing a product that is of good quality, and being in general user-friendly and relatively easy to tune which has kept commissioning times within reasonable bounds. However, as noted previously in Chapter 4, because of their SISO-type structure, the present systems are limited in realizing further improvements in performance due to their limited capability to handle the dynamic interactions between the many variables in the mill itself, so that a control structure which offers an improvement over the conventional arrangements is considered essential for improvement beyond the present limitations.

Several different control structures were considered as possible viable candidates for mill control. While there are certain advantages and disadvantages of each candidate as will be discussed later in this chapter, the one that offers the best chance of improvement in performance, that promotes physical intuition in the design process, and does not add complexity that would be unacceptable to commissioning and operational personnel most of whom who are unfamiliar with advanced control techniques, is the state-dependent Riccati equation method as presented herein. During the investigation of this technique, it was discovered that performance and user-friendliness could be improved even further by the addition of very simple augmentations to the basic SDRE structure. This improvement turned

J. Pittner and M.A. Simaan, *Tandem Cold Metal Rolling Mill Control*,
Advances in Industrial Control, DOI 10.1007/978-0-85729-067-0_5,
© Springer-Verlag London Limited 2011

out to be quite effective in the development of an overall system that offered significant improvements.

The state-dependent Riccati equation technique itself is not new, as it was first suggested by Pearson [1] in the early 1960s, later developed further by others [2] in the mid 1970s, studied further in the late 1990s by Mracek and Cloutier [3], and also referred to by Friedland [4] mid 1990s. It was not until in the past 10 years or so that the technique has been recognized as a practical and user-friendly method for the control of nonlinear systems. This is mostly due to the rapid advancements in the capability of computational machines which made the online pointwise solution of the algebraic Riccati equation feasible for many different types of processes that require fast updates in controller scans, such as tandem cold metal rolling. Many complex nonlinear applications using the basic SDRE method were implemented successfully. Among these are an artificial human pancreas [5], a nonlinear autopilot [6], an advanced guidance law development [7], a satellite and spacecraft control [8], a process reactor control [9], plus others.

In preparation for consideration of the material related to the state-dependent Riccati equation technique, it is necessary to have some basic understanding of control using the linear quadratic regulator (LQR) method, and so accordingly what is included in the following section are some of the basics of the applicable areas of LQR control. As with the material presented throughout this book, what is presented is oriented mostly toward practical usage, rather than from the standpoint of a highly formal mathematical approach, although some of the ideas introduced require a familiarity of certain concepts which are described by various mathematical relationships. As in other chapters, a more mathematically rigorous and detailed coverage can be found in the references and the citations listed therein.

5.2   The Linear Quadratic Regulator

There are several aspects to control using the techniques of linear quadratic regulation. The general idea in all of these aspects is the development of a control law which when applied to a dynamical system will minimize a performance index that represents a cost function. The minimization of this index ultimately results in a controller that is optimal in the sense that when applied to a certain process the controller can inherently reduce the undesired deviations in certain state variables from their desired values during the operation of the controller. In addition to reducing the deviations in these state variables, the cost function can be constructed to limit the control effort (*i.e.* the control energy) needed to perform the control action, and thus minimize the overall cost. For the purposes of what is being considered in this section, the elements of the coefficient matrices of the linearized model and the elements of the weighting matrices are taken to be constants. The determination of the cost function is based on a linearization of the process model, hence the term "linear" in the description of the controller. The cost function is usually a combination of a quadratic function of the state variables and a quadratic function of the control variables hence the term "quadratic". A quadratic cost function is almost always used as it fits very nicely into this method as compared to other possible functions, the resulting optimal control is expressed in the form of a linear feedback, and the resulting closed-loop system is a linear dynamic system.

The elements of the weighting matrices are assigned by the control system designer. By proper assignment of these elements the control designer can "penalize" (*i.e.* emphasize) certain variables of the linearized system to make them more dominant during the action of the controller. The capability to assign these elements also allows for "trade-offs" between weightings assigned to the states versus weightings assigned to control, as for example a faster response in certain of the states can be achieved at the expense of a greater control effort. For the purposes of providing insight into the parallels with the SDRE method, it will be assumed that the LQR control is closed-loop and that the controller operates continuously for an infinite or a very long period of time, as opposed to being open-loop or functioning for a much shorter finite period of time as is the case for certain other applications of the linear quadratic method. A linearized state equation and a performance index J which represents a quadratic cost function are given in (5.1) and (5.2),

$$\frac{dx}{dt} = \dot{x} = Ax + Bu, \quad x(0) = x_0, \tag{5.1}$$

$$J = \frac{1}{2} \int_0^\infty (x'Qx + u'Ru)dt, \tag{5.2}$$

where $x \in R^n$ is the state vector, $u \in R^m$ is the control vector, $A \in R^{n \times n}$, $B \in R^{n \times m}$ are the coefficient matrices, $Q \in R^{n \times n}$, $R \in R^{m \times m}$ are the state and control weighting matrices and J is the performance index which is to be minimized, and $'$ represents the transpose of a matrix.

The control law (5.4) which minimizes J, denoted in this case as an infinite horizon performance index, is determined by the solution of an algebraic Riccati equation (5.3),

$$A'K + KA + KBR^{-1}B'K + Q = 0, \tag{5.3}$$

where the constant matrix $K \in R^{n \times n}$ is the solution to (5.3), and the control law is

$$u = -R^{-1}B'Kx. \tag{5.4}$$

The state equation (5.1) is considered to be a constraint equation which is to be satisfied concurrently with (5.2). To assure a solution to (5.3) the pair (A, B) must be stabilizable. The solution K is unique and is positive semi-definite if there is a matrix C such that $Q = C'C$ and the pair (A, C) is detectable. K is positive definite if (A, C) is observable. Also, Q is to be positive semi-definite and R positive definite (*i.e.*, $Q \geqslant 0$ and $R > 0$). If Q is selected so that $(A, \sqrt{Q})$ is observable then the closed-loop system is assured to be stable. A simple example follows which illustrates the application of the LQR method. More detailed material related to LQR control is available in [10, 11].

In this example, the linearized system is described in (5.5) and (5.6) as

$$\dot{x} = Ax + Bu, \quad x(0) = [-1 \;\; 5]', \tag{5.5}$$

$$y = Cx, \tag{5.6}$$

where

$$A = \begin{bmatrix} 0 & 1 \\ 0 & 0 \end{bmatrix}, \quad B = \begin{bmatrix} 0 \\ 1 \end{bmatrix}, \quad C = \begin{bmatrix} \sqrt{10} & 0 \end{bmatrix}. \tag{5.7}$$

For the LQR controller for this system the elements of Q have been chosen to emphasize x_1, and with R taken as I, so that

$$Q = \begin{bmatrix} 10 & 0 \\ 0 & 0 \end{bmatrix}. \tag{5.8}$$

The resulting algebraic Riccati equation has the unique positive definite solution

$$K = \begin{bmatrix} 7.9527 & 3.1623 \\ 3.1623 & 2.5149 \end{bmatrix}. \tag{5.9}$$

The resulting control law is

$$u = -R^{-1}B'Kx, \quad \text{or} \quad u = -[3.1623 \;\; 2.5149] \; x. \tag{5.10}$$

The optimal feedback system is then

$$\dot{x} = \hat{A}x + Br, \quad y = Cx, \tag{5.11}$$

where

$$\hat{A} = A - BR^{-1}B'K \quad \text{or} \quad \hat{A} = \begin{bmatrix} 0 & 1 \\ -3.1623 & -2.5149 \end{bmatrix}, \tag{5.12}$$

and $r \; (= 0)$ is the system reference. The eigenvalues of $\hat{A}$ are $-1.2574 \pm j1.2574$ which implies that the system is closed-loop stable as the eigenvalues have negative real parts. Figure 5.1 depicts the system showing the optimal feedback.

In this system the controller acts to drive the system states from their initial values $x(0) = [-1 \;\; 5]'$ to zero as depicted in Figure 5.2. Had the reference r been

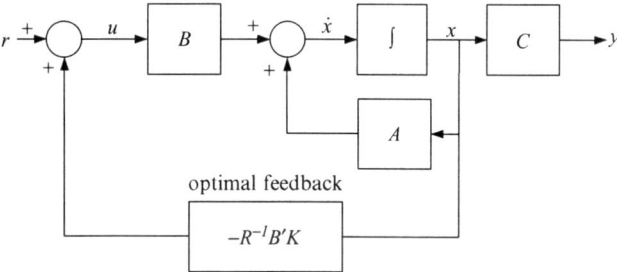

Fig. 5.1 System configuration

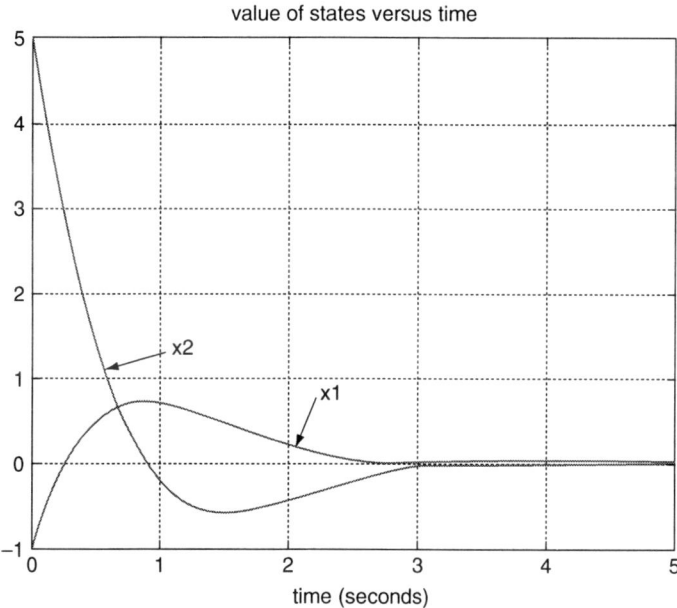

Fig. 5.2 State responses

other than zero, appropriate adjustments would be made to the u and x terms in the cost function (5.2) to account for the shift in the operating point, similarly to what is shown later in this chapter wherein the SDRE control is considered.

5.3 The Augmented SDRE Controller

The basic SDRE controller has the same structure as the LQR controller except that the A, B, and C coefficient matrices in the state and output equations and the Q and R weighting matrices in the cost function are state-dependent. While at first glance this may seem to be a fairly simple revision from a theoretical standpoint, a bit more reflection will reveal that this certainly is not the case as will be seen in the following sections. In fact using analytical methods for the estimation of stability and performance, which includes robustness to uncertainties and disturbances, is quite difficult and actually not possible in most cases. This is because, as of the writing of this book, there has yet to be a comprehensive supporting theoretical background for the application of this technique, although there are some results as described in what follows. As a result, the application of the SDRE method requires the reliance on simulation to verify stability and performance. However, the dearth of theoretical support has not impeded progress in the application of this very useful method and many highly successful results have been realized when applying the basic SDRE technique to a variety of processes, several of which have been noted earlier, so that it is becoming recognized as a highly useful nonlinear method for the control of nonlinear systems which offers several desirable features that are generally not available in other nonlinear control methods.

5.3.1 *Theoretical Background*

While there is an overall absence of supporting theory for SDRE, there are some results that are reported which can provide some insight into the existing theory. These are presented in this section and in its Appendix which lists several applicable definitions and theorems. More specific detail regarding the theoretical background for SDRE is available in [12–14].

In the theoretical treatment it is assumed that the dynamics of a nonlinear plant can be described by a state equation (5.13) and an output equation (5.14),

$$\frac{dx}{dt} \equiv \dot{x} = a(x) + b(x)u, \tag{5.13}$$

$$y = g(x), \tag{5.14}$$

where $x \in R^n$ is a vector whose elements represent the individual state variables, $a(x) \in R^n$ is a state-dependent vector, $u \in R^m$ is a vector whose elements represent the individual control variables, $y \in R^p$ is a vector whose elements represent the individual output variables, $b(x) \in R^{n \times m}$ is a state-dependent matrix, $g(x) \in R^p$ is a state-dependent vector, with $a(0) = 0$, $g(0) = 0$, $a(x) = C^1$, and $g(x) = C^1$. By factorizing[1] the state-dependent vectors $a(x)$ into $A(x)x$, $g(x)$ into $C(x)x$, and with $b(x) = B$, the above becomes a form resembling linear state space equations,

---

[1]It is known [16] that if $a(0) = 0$, and $a(x) \in C^1$, an infinite number of such factorizations exist, and similarly with $g(x)$.

$$\dot{x} = A(x)x + Bu, \quad x(0) = x_0, \tag{5.15}$$

$$y = C(x)x, \tag{5.16}$$

where $A(x) \in R^{nxn}$ is a state-dependent matrix, $C(x) \in R^{pxn}$ is a state-dependent matrix, $B(\neq 0) \in R^{nxm}$ is a constant matrix, with x, u, and y as noted previously. Similar to the LQR problem the optimal control problem in the SDRE method is to minimize the performance index

$$J = \frac{1}{2} \int_0^\infty (x'Q(x)x + u'R(x)u)dt, \tag{5.17}$$

with respect to the control vector u, subject to the constraint (5.15), where $Q(x)$ and $R(x)$ are state-dependent weighting matrices, with $Q(x) \geq 0$, $R(x) > 0$, and $a(x)$, $Q(x)$, $R(x) \in C^k$, $k \geq 1$, for all x in the control space.[2] The objective of (5.17) is to find a control law which regulates the system to the origin.

The method of solution is first to find a factorization of $a(x)$ such that (5.13) can be expressed in the form of (5.15). Then, similar to the LQR method, the state-dependent algebraic Riccati equation

$$A'(x)K(x) + K(x)A(x) - K(x)BR^{-1}(x)B'K(x) + Q(x) = 0, \tag{5.18}$$

is solved pointwise for $K(x)$ resulting in the control law

$$u = -R^{-1}(x)B'K(x)x, \tag{5.19}$$

where $K'(x) = K(x) \geq 0$, for all x. In order to insure a solution to (5.18) at each point, the method requires that the pair $(A(x), B)$ be pointwise stabilizable for all x.

Asymptotic stability for all x must be confirmed by simulation as there is no useful theory which assures it, except for certain special cases [13]. Local asymptotic stability is assured by theorem as developed in [15] and as noted in the Appendix. It also should be noted that, even though the SDRE technique produces a closed-loop system matrix which has its eigenvalues in the open left-half plane for all x, this does not assure global asymptotic stability. This is shown by the following interesting example [17].

Considering (5.20) which represents the closed-loop dynamics of a plant coupled to an SDRE controller with control law (5.19),

$$\dot{x} = a_{cl}(x), \quad x_1(0) = 2, \quad x_2(0) = 2, \tag{5.20}$$

---

[2]Unless noted otherwise, when the expression "for all x" is used, it is intended to mean "for all x in the control space".

where $x \in R^2$ is the state vector, $a_{cl}(x) \in R^2$ is a state-dependent vector, x_1 and x_2 are the elements of x, with

$$a_{cl,1}(x) = -x_1 + x_1^2 x_2, \tag{5.21}$$

$$a_{cl,2}(x) = -x_2. \tag{5.22}$$

Factorizing (5.21) in the form $a_{cl}(x) = A_{cl}(x)x$ gives

$$A_{cl}(x) = \begin{bmatrix} -1 & x_1^2 \\ 0 & -1 \end{bmatrix}. \tag{5.23}$$

The eigenvalues of $A_{cl}(x)$ are $(-1, -1)$ for all x. The solution to (5.20) is

$$x_1(t) = \frac{2x_2(t)}{x_2^2(t) - 2}, \tag{5.24}$$

$$x_2(t) = 2e^{-t}. \tag{5.25}$$

As easily can be seen, as $t \to log\sqrt{2}$, $x \to \infty$, which implies that the system is unstable even though all eigenvalues lie in the open left-half plane.

The optimality of systems controlled by the SDRE technique method has been examined and a necessary condition for optimality has been derived using the Lagrange multiplier technique for the case where $b(x) \in R^{nxm}$ in (5.13) is a state-dependent matrix. For the case of interest herein where $b(x)$ is a constant matrix $B \in R^{nxm}$ the derivation is similar. In the Lagrange multiplier method which is based on Lagrange theory using the calculus of variations, a Hamiltonian function is formed from the cost function (5.17) and the nonlinear constraint (5.15) as

$$H(x, u, \lambda) = \tfrac{1}{2}(x'Q(x)x + u'R(x)u) + \lambda'(A(x)x + Bu), \tag{5.26}$$

where $\lambda \in R^n$ is a Lagrange multiplier. Using the Hamiltonian function, the necessary conditions for optimality of a nonlinear controller can be derived in the form of a state equation (5.27), a costate equation (5.28), and a stationarity condition (5.29),

$$\nabla_\lambda H = \dot{x}, \tag{5.27}$$

$$\nabla_x H = -\dot{\lambda}, \tag{5.28}$$

$$\nabla_u H = 0. \tag{5.29}$$

Solving (5.27), (5.28), and (5.29) as in the Appendix results in a necessary condition (5.30) for the optimality of the SDRE controller, *i.e.* if this condition is satisfied, the necessary conditions for optimality of the nonlinear controller also are satisfied,

$$\dot{K}(x)x + \tfrac{1}{2}(x'\nabla_x Q(x)x + x'K(x)BR^{-1}(x)\nabla_x R(x)R^{-1}(x)B'K(x)x)$$
$$+ x'\nabla_x A'(x)K(x)x = 0, \tag{5.30}$$

where the computation of $x'\nabla_x Q(x)x$ and other gradient functions of (5.30) are as described in the Appendix. In general this condition is satisfied only for a unique $A(x)$, which is difficult to determine unless the optimal cost function is known a priori. However, if the matrix functions $A(x)$, $K(x)$, $Q(x)$, and $R(x)$, and their gradients $\nabla_x A(x), \nabla_x K(x), \nabla_x Q(x)$, and $\nabla_x R(x)$ are bounded for all x, and under global asymptotic stability, then as shown in [15] the state trajectories converge to the optimal state trajectories as the states are driven to zero. This is taken to be a near-optimal (*i.e.* a suboptimal) condition in a neighborhood of the origin.

As can be seen from the foregoing the useful theory is scant for assuring global asymptotic stability and for predicting performance, including robustness to uncertainties and disturbances, and therefore concerns regarding these issues must be resolved on a case basis by simulation. Additionally, in the case of the tandem cold rolling process, the process itself is quite large, is highly nonlinear with significant time delays that vary considerably with the mill speed, and there are major uncertainties and disturbances that must be mitigated by the controller. Moreover, there is the requirement that the controller be user-friendly, which implies that it promotes the use of physical intuition in the design process and is easy to tune for commission personnel who have limited backgrounds in advanced control theory. The next section considers how these issues are addressed in the augmented SDRE technique.

5.3.2 Considerations for the Application to Tandem Cold Rolling

As has been previously mentioned, the characteristics of the tandem cold rolling process that make it an especially challenging control application are its large size, its nonlinearity, the multiple and significant time delays whose values change considerably with the mill speed, the major disturbances plus a broad range of uncertainties both of which change with the operating point, with the product being processed, and with the changes in the mill characteristics that occur during normal processing operations. Additionally there is the need for the controller to be user-friendly to design and commissioning personnel. In this section it is shown how a controller using the basic SDRE method with augmentations handles these requirements.

In tandem cold rolling generally the operating point of the process is established by the production schedule which is based on the capabilities of the mill to process a particular product. It is the function of the mill controller for centerline thicknesses and tensions to reduce excursions in these variables from their desired values at the operating point during the various regimes of mill operation and in the presence of disturbances and uncertainties. In the case of the SDRE method, the performance index is modified to account for a non-zero operating point by the introduction of a variable z which changes the coordinates to correspond to a particular operating point,

$$z = x - x_{op}, \tag{5.31}$$

where $z \in R^n$, $x \in R^n$ is the state vector, with $x_{op} \in R^n$ being a vector whose elements represent the values of the individual state variables at the operating point. The cost function then is modified to be

$$J = \tfrac{1}{2} \int_0^\infty \left(z'Qz + (u - u_{op})'R(u - u_{op}) \right) dt, \tag{5.32}$$

where $u \in R^m$ is the control vector, and $u_{op} \in R^m$ is a vector whose elements represent the values of the individual control variables at the operating point. For simplification the weighting matrices are taken initially as diagonal matrices with tunable constant elements. The operating point will change with the mill speed and the product being processed. For a given product, certain elements of x_{op} and u_{op} are changed during a speed change, as shown in Table 5.1, where the variable assignments for the state, control, and output vectors of Table 2.1 are repeated in Table 5.2 for the reader's convenience. The remaining elements of x_{op} and u_{op} are taken for this evaluation as remaining unchanged during the speed change.

The controller must mitigate the effects of both external and internal disturbances in the presence of uncertainties. As initially described in Section 3.2, the external disturbances include changes in the strip thickness and the strip hardness at the mill entry, both of which arise mostly from previous processing in the hot

Table 5.1 Operating point variables associated with x_{op} and u_{op} that are changed during a mill speed change

Variables associated with	
x_{op}	u_{op}
$x_{10,op}$ $(V_{1,op})$	$u_{6,op}$ $(U_{V1,op})$
$x_{11,op}$ $(V_{2,op})$	$u_{7,op}$ $(U_{V2,op})$
$x_{12,op}$ $(V_{3,op})$	$u_{8,op}$ $(U_{V3,op})$
$x_{13,op}$ $(V_{4,op})$	$u_{9,op}$ $(U_{V4,op})$
$x_{14,op}$ $(V_{5,op})$	$u_{10,op}$ $(U_{V5,op})$

Table 5.2 Variable assignments for state, control, and output vectors (Table 2.1)

State vector		Control vector		Output vector	
x_1 (σ_{12})	x_8 (S_4)	u_1 (U_{S1})	u_6 (U_{V1})	y_1 (h_{out1})	y_8 (σ_{34})
x_2 (σ_{23})	x_9 (S_5)	u_2 (U_{S2})	u_7 (U_{V2})	y_2 (h_{out2})	y_9 (σ_{45})
x_3 (σ_{34})	x_{10} (V_1)	u_3 (U_{S3})	u_8 (U_{V3})	y_3 (h_{out3})	y_{10} (P_1)
x_4 (σ_{45})	x_{11} (V_2)	u_4 (U_{S4})	u_9 (U_{V4})	y_4 (h_{out4})	y_{11} (P_2)
x_5 (S_1)	x_{12} (V_3)	u_5 (U_{S5})	u_{10} (U_{V5})	y_5 (h_{out5})	y_{12} (P_3)
x_6 (S_2)	x_{13} (V_4)			y_6 (σ_{12})	y_{13} (P_4)
x_7 (S_3)	x_{14} (V_5)			y_7 (σ_{23})	y_{14} (P_5)

rolling area. In instances where the slabs originate in a reheating furnace, certain changes in thickness and hardness are caused by colder areas in the slabs prior to hot rolling. These colder areas result from contact of the hot metal with the support skids in the reheat furnace, and occur even though this effect (referred to as skid chill) is reduced considerably by rocking the slabs on the skids. Thus when processed in the hot mill, the slabs have gradients in temperature which cause variations in thickness and hardness to be rolled into the strip. These variations are approximately periodic, with thickness and hardness changing essentially in phase with each other.

In the case of mill entry thickness, disturbances also result from the eccentricity effects of the hot mill rolls. The so-called roll eccentricity is actually a combination of effects (Section 3.5) which results in a cyclic variation in the strip thickness during rolling. The frequency of this variation is much higher than that of the variation in thickness caused by the skid chill effect as can be seen in Figures 3.1 and 3.2 which depict typical disturbances in entry thickness and in entry hardness at a typical run speed and at a typical thread speed. Internal disturbances are eccentricity effects of the cold mill rolls which are addressed separately in Section 5.3.4.

The major uncertainties are those in modeling and in measurement as described in Section 3.2. Table 3.1 lists typical modeling uncertainties and Table 3.2 lists typical uncertainties in measurement. Uncertainties in the model and in the measurements of process variables can result in undesirable deviations in the stand output thicknesses and in the interstand tensions from their values at an operating point, and thus a strong robustness to these uncertainties is essential.

Both disturbances and uncertainties are modeled as what they are and where they are. Their combined effect is represented as a change in the elements of the $A(x)$ matrix and a change in the elements of the $C(x)$ matrix, so that (5.15) and (5.16) become

$$\dot{x} = A(x)x + \delta A(x)x + Bu, \quad x(0) = x_0, \tag{5.33}$$

$$y = C(x)x + \delta C(x)x, \tag{5.34}$$

where $\delta A(x) \in R^{n \times n}$ is a matrix whose elements are changes in the elements of $A(x)$ and similarly for $\delta C(x) \in R^{p \times n}$.

5.3.3 Controller Configuration

During the initial investigation [18] of the SDRE method for control of the tandem cold mill, consideration was given to using the basic SDRE structure without augmentations. A brief review of this initial investigation is included to provide some insight into the functioning of the augmentations and how they contribute to the final control of the system.

5.3.3.1   Initial System Configuration

The initial system configuration which is depicted in Figure 5.3, is based on the use
of the modified state and output equations (5.33) and (5.34) with the control law
revised to include the variable z (5.31) which effects a coordinate shift to account
for the non-zero operating point,

$$u = -R^{-1}B'K(x)z. \tag{5.35}$$

In this simulation the elements of matrices $A(x)$, B, and $C(x)$ are determined as
noted in Section 5.3.3.5, with the weighting matrices Q and R initially taken to be
diagonal with tunable constant elements, so that the algebraic Riccati equation is
modified to be

$$A'(x)K(x) + K(x)A(x) - K(x)BR^{-1}B'K(x) + Q = 0, \tag{5.36}$$

where the state-dependency of the solution $K(x)$ is retained. In the initial system
depicted in Figure 5.3 x_{op} represents the states at the operating point and $u_f \in R^m$ is
a vector whose elements represent the variables computed by the control law (5.35).
For this initial investigation it was assumed that the algebraic Riccati equation
(ARE) could be solved fast enough to properly control the mill. The actual method
of solving the ARE sufficiently fast is addressed in Section 5.3.3.7.

The mitigation of the effects of disturbances in the incoming strip thickness and
hardness is an important function of the controller. To determine the capability for

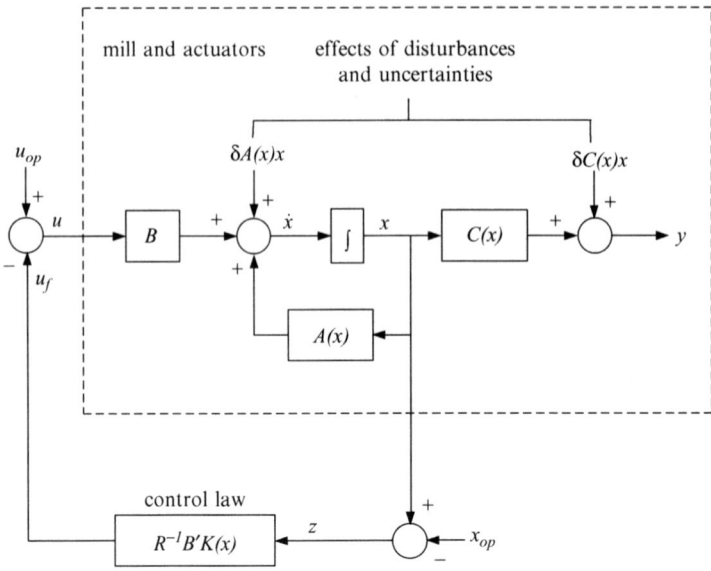

Fig. 5.3 Initial system configuration without augmentations

the controller depicted in Figure 3.3 to perform this function, an initial simulation was performed with the disturbance in mill entry thickness as shown in Figure 3.1 applied with the mill at run speed (1,220 m/min). For this initial simulation the interstand time delays were taken to be zero. With no other disturbances or uncertainties applied several adjustments in the elements of the matrices Q and R were made to attempt to reduce the magnitude of the peak excursion in the centerline thickness at the exit of stand 1. The lowest peak excursion that could be realized was about 2%, which was considered unacceptable. Additional states then were added to the model to approximate the interstand time delay as a series of four first order lags, and the simulation repeated. With adjustments made in the Q and R matrices under the same conditions as previously, the magnitude of the excursions was much better, about 0.2% which was mostly attributed to the setting of the appropriate diagonal element in the weighting matrix Q to penalize the added state that represented the input thickness to stand 2. While this was a significant improvement, it required the addition of 16 states for the entire mill. Additionally, the error in using this approximation to the time delay was considered less than acceptable. Using other methods which could improve the approximation of the time delay would add even more states.

The initial simulation with the time delays at zero was repeated, except with no disturbances and with the uncertainties listed in Table 5.3, each applied separately and then concurrently in a manner to simulate more severe conditions. The results as summarized in Table 5.3 also are unacceptable and are expected to be worse during changes in mill speed.

The addition of 16 or more states is unacceptable for mitigating disturbances and uncertainties, as is the addition of an outer loop thickness trim using a BISRA method to estimate the output thicknesses. This is due to the uncertainty in the mill modulus plus the effects of eccentricity in the mill rolls. An interim solution is the addition of very high accuracy (uncertainty of $+/-0.025\%$) laser-based strip speed sensors at the entry of mill stands 2, 3, 4, 5 and at the exit of stand 5 to estimate the output thickness using mass flow techniques, with a subsequent outer loop trim based on the thickness estimate. This type of high quality speed sensor is in use successfully at several installations and has proven reliability but requires additional maintenance efforts. However, the additional maintenance is considered acceptable if a significant improvement in the quality of the final product can be realized. Further simulations as in Section 5.4 have confirmed this improvement in quality.

Table 5.3 Effects of modeling uncertainties on stand 1 output thickness

Variable	Uncertainty	Magnitude of maximum percent change in stand 1 output thickness
μ	20%	0.7%
k	25	4.5
M	-10	2.6
μ, k, M (concurrently)		6.0

With the speed sensors added and the closed-loop trims configured, initial simulations were performed with the time delays simulated using the variable time delay function in Simulink[3] and with the outer loop thickness trim enabled. These initial simulations were quite successful as almost negligible excursion in the output thickness was observed. This is because all of the more significant uncertainties are inside the closed loops of the trims, and therefore are mitigated by the closed-loop control action. By setting the appropriate elements of the Q and R matrices a stronger emphasis then was given to the states representing interstand tensions which significantly reduced the excursions in the tensions. A slightly further reduction in the tension excursions was achieved by the addition of a second outer loop trim on the tension reference, which also offered some additional advantages which are described later in this chapter.

5.3.3.2 Basic Controller Structure

Based on the results of these initial simulations a basic structure of the controller (Figure 5.5) was established for further evaluation. Figure 5.4 depicts the configuration of a five-stand stand-alone mill with the strip speed sensors added.

The added trims for the thickness and the tension are as shown as in Figure 5.5. In Figure 5.5 y_m represents the measureable elements of the output vector y, $V_{in,i}$ ($i = 2,3,4,5$) are the measured strip speeds at the inputs of stands 2, 3, 4, and 5, V_{out5} is the measured strip speed at the output of stand 5, h_{out1m} and h_{out5m} are the measured strip thicknesses at stand 1 and stand 5, $y_e \in R^p$ ($p = 14$) is a vector whose elements are estimates of the elements of y, ϕ_y is an algorithm which uses h_{out1m}, h_{out5m}, $V_{in,i}$, V_{out5}, y_m, and the measured values of certain variables represented by the elements of the state vector x to generate y_e, $K_I \in R^{m \times p}$ ($m = 10, p = 14$) and

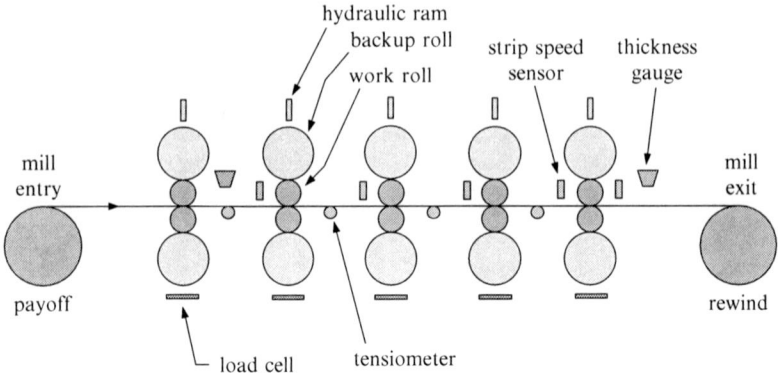

Fig. 5.4 Five stand stand-alone tandem cold mill with added strip speed sensors

---

[3]Simulink is a registered trademarks of The MathWorks, Inc., Natick, MA 01760-2098.

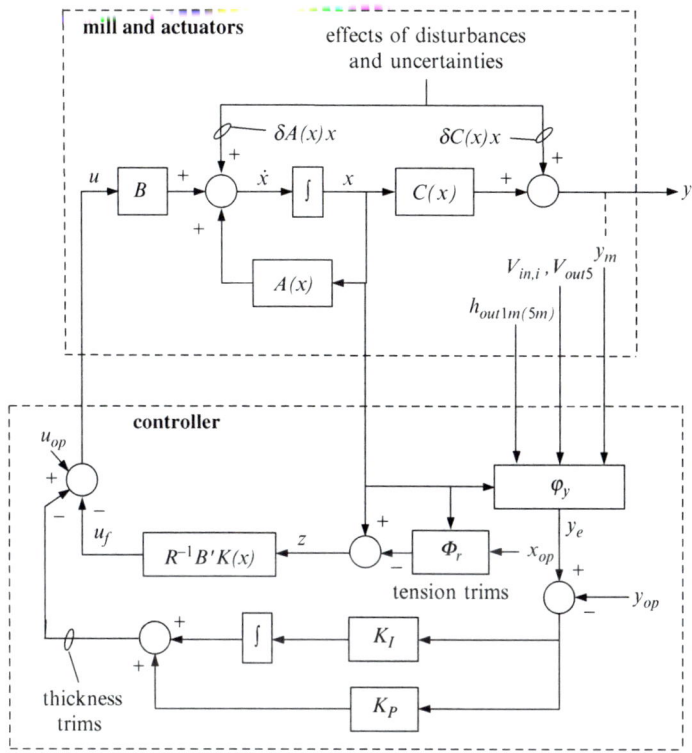

Fig. 5.5 System configuration with trims as augmentations

$K_p \in R^{m \times p}$ are matrices whose elements are zero except for elements (j,j), $(j = 1,2,3,4,5)$, which are the gains for the integral and proportional thickness trim functions for each stand.

5.3.3.3 Strip Thickness Estimation

Under steady-state conditions and with almost constant strip width, the strip thickness at the output of a mill stand can be estimated using conservation of mass flow across the roll gap as

$$h_{out,i} = \frac{h_{in,i} V_{in,i}}{V_{out,i}} k_{i,e}, \quad (i = 1,2,3,4,5), \tag{5.37}$$

where $k_{i,e}$ is a correction factor to adjust for small errors in the estimated output thickness caused by spreading of the strip, reduction in the strip width, and other effects. Substituting the strip speed at the input of the next stand for the output strip speed gives, for stands 2,3, and 4,

$$h_{out,i} = \frac{h_{in,i} V_{in,i}}{V_{in,i+1}} k_{i,e}, \quad (i = 2, 3, 4), \tag{5.38}$$

with (5.37) and $i = 5$ being used for stand 5. This substitution, which eliminates the need for a speed sensor at the output of the stands 2, 3, and 4, results in very little error in the estimation of the output thickness as shown by simulation in Section 5.4.2. In addition, the simulation also has confirmed that using the relationship (5.38) during transient conditions does not cause significant error in the estimation of the output thickness. During operation, a separate mill adaptation system determines the correction factors $k_{i,e}$. For example, a possible method for setting k_{5e} would be to compare the uncompensated mass flow computed thickness

$$h_{out5mf} = \frac{V_{in5}}{V_{out5}} h_{in5e}, \tag{5.39}$$

as tracked from the mill stand to the thickness gauge, against the measured thickness h_{out5m}, and then use smoothing functions based on the previous estimates of k_{5e}. The adjustment for stand i ($i = 2,3,4$), is estimated by using the adjustment determined for stand 5 as modified by the scheduled input/output thickness ratios for stand i. The effects of $k_{i,e}$ on the error in the mill exit thickness are considered in Section 5.4.5.

The physical location of the thickness gauge at the exit of stand 1 causes a time delay in the measured thickness of the strip exiting the stand, and similarly for stand 5. Values of this time delay at a typical run speed and at a typical thread speed are given in Table 5.4 for a typical gauge location of 1 m from the stand.

Because of the time delay in the thickness measurement, faster transient errors in thickness at the exit of the stand will result in greater deviations in thickness from the operating point value than if there were no time delay. To partially compensate for this at stand 1, a BISRA estimate (3.38) of the thickness at the exit of the stand is made, which then is trimmed by the difference between the measured thickness and a previous BISRA estimate delayed by the time delay, as depicted in Figure 5.6. The integral gain K_{g_int1} is set intuitively and is confirmed by simulation. Thus there is some immediate correction for faster transient errors while the steady-state thickness is held to the desired operating point value.

As part of the algorithm ϕ_y the BISRA estimate is

$$h_{out1b} = x_5 + S_0 + \frac{F_{1m}}{M_{1e}}, \tag{5.40}$$

Table 5.4 Values of time delay from stand 1(5) to the downstream thickness gauge

Strip speed	Time delay from stand to thickness gauge	
	Stand 1	Stand 5
run, 1,220 m/min	0.10 s	0.05
thread, 61	1.90	0.90

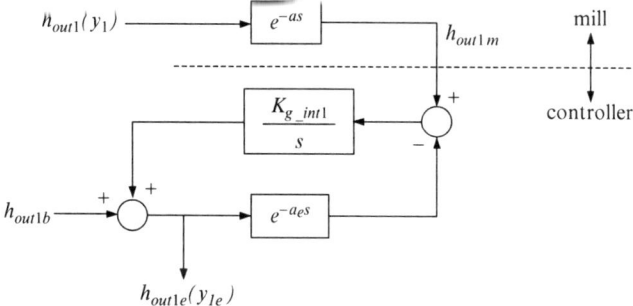

Fig. 5.6 Stand 1 output thickness estimation

where x_5 is the stand 1 position actuator position, S_0 is the intercept of the linearized portion of the mill stretch curve for stand 1, F_{1m} is the measured total roll force for stand 1, and M_{1e} is the estimated mill modulus for stand 1. The effects of backup roll eccentricity are addressed in Section 5.3.4. In Figure 5.6, the notation $h_{out1e}(y_{1e})$ indicates that the variable h_{out1e} is represented by element 1 of vector y_e, and similarly for other variables represented by the elements of y and y_e. In Figure 5.6, the time delay from stand 1 to the thickness gauge is represented as a. The tracking of thickness in the controller from stand 1 to the thickness gauge is represented as a time delay

$$a_e = \frac{L_{m1}}{V_{in2}}, \tag{5.41}$$

where L_{m1} is the distance from stand 1 to the thickness gauge, and V_{in2} is the measured strip speed at the input of stand 2, which very closely approximates the strip speed at the output of stand 1 and is used for the tracking of the thickness. As noted previously, the BISRA estimate is very sensitive to uncertainties in the estimated mill modulus M_{1e}. To reduce the effects of this uncertainty, M_{1e} is estimated by applying a BISRA relationship for an element of strip at the thickness gauge as

$$h_{out1m}(t) = x_5(t - a_e) + S_0 + \frac{F_{1m}(t - a_e)}{M_1(t - a_e)}, \tag{5.42}$$

or by rearranging

$$M_1(t - a_e) = \frac{F_{1m}(t - a_e)}{h_{out1m}(t) - x_5(t - a_e) - S_0}, \tag{5.43}$$

and approximating $M_{1e}(t)$ as

$$M_{1e}(t) \cong M_1(t - a_e), \tag{5.44}$$

where $F_{1m}(t - a_e)$, $x_5(t - a_e)$, and $M_1(t - a_e)$ are the variables $F_{1m}(t)$, $x_5(t)$, and $M_1(t)$ delayed by a_e. As noted earlier, changes in M_1 are mostly the result of changes in the backup roll diameter due to mechanical wear and heating effects, which occur slowly compared to the time delay a_e. Thus it is reasonable to approximate $M_{1e}(t)$ by $M_1(t - a_e)$.

The measured thickness h_{out1m} is tracked in the controller from the thickness gauge at the exit of stand 1 to stand 2 using the measured strip speed V_{in2}, so that the input thickness at stand 2 is determined as

$$h_{in2e}(t) = h_{out1m}(t - (\tau_{d12} - a_e)), \tag{5.45}$$

where τ_{d12} is the estimated time delay (2.28) between stand 1 and stand 2. Then using (5.38) the output thickness at stand 2 is estimated in the controller as

$$h_{out2e} = \frac{h_{in2e}V_{in2}}{V_{in3}}k_{2e}. \tag{5.46}$$

Similarly the thickness at the input to stand 3 can be estimated in the controller by using the measured strip speed at the input of stand 3 to track the thickness from stand 2 to stand 3, so that the input thickness at stand 3 is approximated by

$$h_{in3e}(t) = h_{out2e}(t - \tau_{d23}), \tag{5.47}$$

where τ_{d23} is the interstand time delay (2.28) between stand 2 and stand 3. Using (5.38) to estimate the stand 3 output thickness gives

$$h_{out3e} = \frac{h_{in3e}V_{in3}}{V_{in4}}k_{3e}. \tag{5.48}$$

The output thickness at stand 4 is estimated in a similar manner. The output thickness at stand 5 is estimated as in (5.37) with $i = 5$, where V_{in5} and V_{out5} are measured variables. In the case of stand 5, both the estimate of the output thickness h_{out5b} and the thickness measurement just downstream of the stand are used in a configuration similar to that of stand 1, as depicted in Figure 5.7, to obtain the thickness estimate $h_{out5e}(y_{5e})$, where h_{out5b} is computed as

$$h_{out5b} = \frac{h_{in5e}V_{in5}}{V_{out5}}k_{5e}, \tag{5.49}$$

with h_{in5e} approximated by

$$h_{in5e}(t) = h_{out4e}(t - \tau_{d45}), \tag{5.50}$$

and where $\tau_{d\,45}$ is the interstand time delay between stand 4 and stand 5.

In Figure 5 7 the time delay fiom stand 5 to the thickness gauge is represented as b. The tracking of thickness in the controller from stand 5 to the thickness gauge is represented as a time delay

$$b_e = \frac{L_{m5}}{V_{out5}},\tag{5.51}$$

where L_{m5} is the distance from stand 5 to the thickness gauge, and V_{out5} is the measured strip speed at the output of stand 5 and is used for the tracking of the thickness. As in the case of stand 1, the integral gain K_{g_int5} is set intuitively and is confirmed by simulation.

5.3.3.4 Interstand Tension Trims

The elements of the Q and R matrices are set to reduce excursions in the interstand tensions. These excursions are reduced further by the addition of operating point trim functions as depicted in Figure 5.8.

In Figure 5.8 $x_{op,i}$ ($i = 1,2,3,4$) is an element of the vector x_{op} which represents the operating point for the interstand tension for stands $i,i + 1$. The reference for the interstand tension for stands $i,i + 1$ is $\sigma_{i,i+1,ref}$, x_i is the is the element of the state vector which represents the measured interstand tension for stands $i,i + 1$, and $K_{i,i+1}$ is a gain term for these stands which is set intuitively and confirmed by simulation. The system configuration (Figure 5.5) includes the algorithm ϕ_r which implements

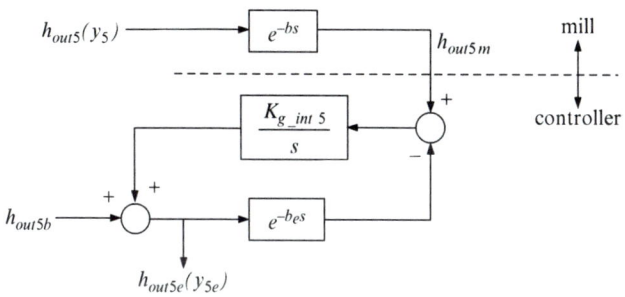

Fig. 5.7 Stand 5 output thickness estimation

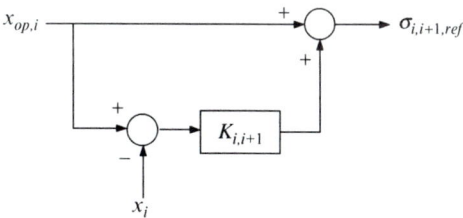

Fig. 5.8 Operating point trim for interstand tension

the interstand tension operating point trims for elements $x_{op,i}$ $(i = 1,2,3,4)$ of the x_{op} vector, and provides a direct feed-through for elements $x_{op,i}$ $(i = 5,\ldots 14)$ of the x_{op} vector. The capability of the operating point trims to provide significant reductions in the excursions in the interstand tensions is shown in the simulation of Section 5.4.

5.3.3.5 Determination of the Elements of the $A(x)$, B, and $C(x)$ Matrices

The matrix $A(x) \in R^{nxn}(n = 14)$ results from a nonunique factorization of $a(x)$ in (5.15) into $A(x)x$. The property of nonunique factorization allows for design flexibility in determining the elements of $A(x)$, which is a desirable feature of the SDRE method. Moreover, the matrix $A(x)$ can be parameterized into $A(x, \alpha)$ where α is a vector of free design parameters, which adds even greater flexibility. In the case of the modeling of the tandem cold rolling process, the vector α was unnecessary and a simple representation $A(x)$ was used [19]. The resulting nonzero elements of $A(x)$ are as shown in Table 5.5 where the notation for the variables used is per Chapter 2. In Table 5.5 the variables $V_{out,i}$ and $V_{in,i}$ can be shown to be functions of x as described in the Appendix.

The nonzero elements of the constant matrix $B \in R^{nxm}(n = 14, m = 10)$ are shown in Table 5.6.

The nonzero elements of the matrix $C(x) \in R^{pxn}(p = 14, n = 14)$ are shown in Table 5.7.

Table 5.5 Expressions for nonzero elements of the $A(x)$ matrix

$A_{1,10} = -\frac{EV_{out1}}{L_0 x_{10}}$	$A_{4,14} = -\frac{EV_{in5}}{L_0 x_{14}}$	$A_{10,10} = -\frac{1}{\tau_V}$
$A_{1,11} = \frac{EV_{in2}}{L_0 x_{11}}$	$A_{5,5} = -\frac{1}{\tau_S}$	$A_{11,11} = -\frac{1}{\tau_V}$
$A_{2,11} = -\frac{EV_{out2}}{L_0 x_{11}}$	$A_{6,6} = -\frac{1}{\tau_S}$	$A_{12,12} = -\frac{1}{\tau_V}$
$A_{2,12} = \frac{EV_{in3}}{L_0 x_{12}}$	$A_{7,7} = -\frac{1}{\tau_S}$	$A_{13,13} = -\frac{1}{\tau_V}$
$A_{3,12} = -\frac{EV_{out3}}{L_0 x_{12}}$	$A_{8,8} = -\frac{1}{\tau_S}$	$A_{14,14} = -\frac{1}{\tau_V}$
$A_{3,13} = \frac{EV_{in4}}{L_0 x_{13}}$	$A_{9,9} = -\frac{1}{\tau_S}$	
$A_{4,13} = -\frac{EV_{out4}}{L_0 x_{13}}$		

Table 5.6 Expressions for nonzero elements of the B matrix

$B_{5,1} = \frac{1}{\tau_S}$		$B_{10,6} = \frac{1}{\tau_V}$
$B_{6,2} = \frac{1}{\tau_S}$		$B_{11,7} = \frac{1}{\tau_V}$
$B_{7,3} = \frac{1}{\tau_S}$		$B_{12,8} = \frac{1}{\tau_V}$
$B_{8,4} = \frac{1}{\tau_S}$		$B_{13,9} = \frac{1}{\tau_V}$
$B_{9,5} = \frac{1}{\tau_S}$		$B_{14,10} = \frac{1}{\tau_V}$

Table 5.7 Expressions for nonzero elements of the $C(x)$ matrix

$C_{1,1} = \frac{h_{out1}}{x_1}$	$C_{6,6} = 1$	$C_{10,10} = \frac{P_1}{x_{10}}$
$C_{2,2} = \frac{h_{out2}}{x_2}$	$C_{7,7} = 1$	$C_{11,11} = \frac{P_2}{x_{11}}$
$C_{3,3} = \frac{h_{out3}}{x_3}$	$C_{8,8} = 1$	$C_{12,12} = \frac{P_3}{x_{12}}$
$C_{4,4} = \frac{h_{out4}}{x_4}$	$C_{9,9} = 1$	$C_{13,13} = \frac{P_4}{x_{13}}$
$C_{5,5} = \frac{h_{out5}}{x_5}$		$C_{14,14} = \frac{P_5}{x_{14}}$

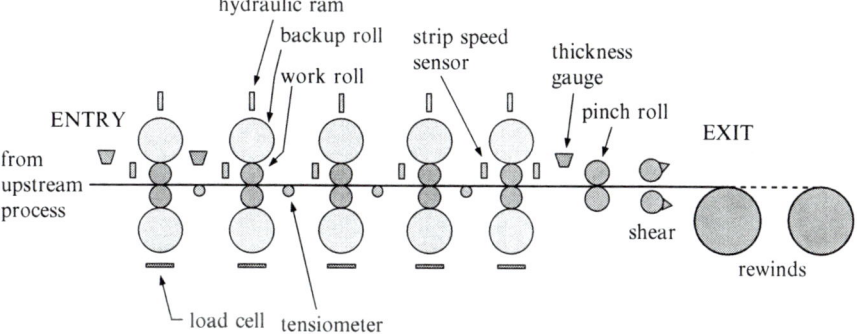

Fig. 5.9 Five-stand continuous tandem cold mill with added strip speed sensors

5.3.3.6 Application to Continuous Mills

The improvements realized in the use of the augmented SDRE technique for control of the stand-alone mill carry over well to the control of the continuous mill. The requirements described in Chapter 3 for control during the change of product on the fly using conventional methods also are applicable for control using the augmented SDRE method. In the case of the continuous mill high accuracy speed sensors are added as in the stand-alone case, except with an additional strip speed sensor located at the entry of the mill for tracking of the strip thickness from the thickness measurement at the mill entry to the first stand, and for weld tracking. Figure 5.9 depicts the configuration of the continuous mill with the added strip speed sensors.

The control for the stand-alone mill previously described using the augmented SDRE method is modified to add logic for reference switching as shown in Figure 5.10. The controller references during the transition from one coil to the next are switched essentially as described in Chapter 3. Section 5.4.4 presents the results of simulations which demonstrate that the improved controller performance of the SDRE technique when applied to the stand-alone mill also carries over well to the continuous mill.

5.3.3.7 Considerations for Actual Usage

In an actual usage, generally the controller is discretized and implemented in a manner to be user-friendly to commissioning personnel whose background usually

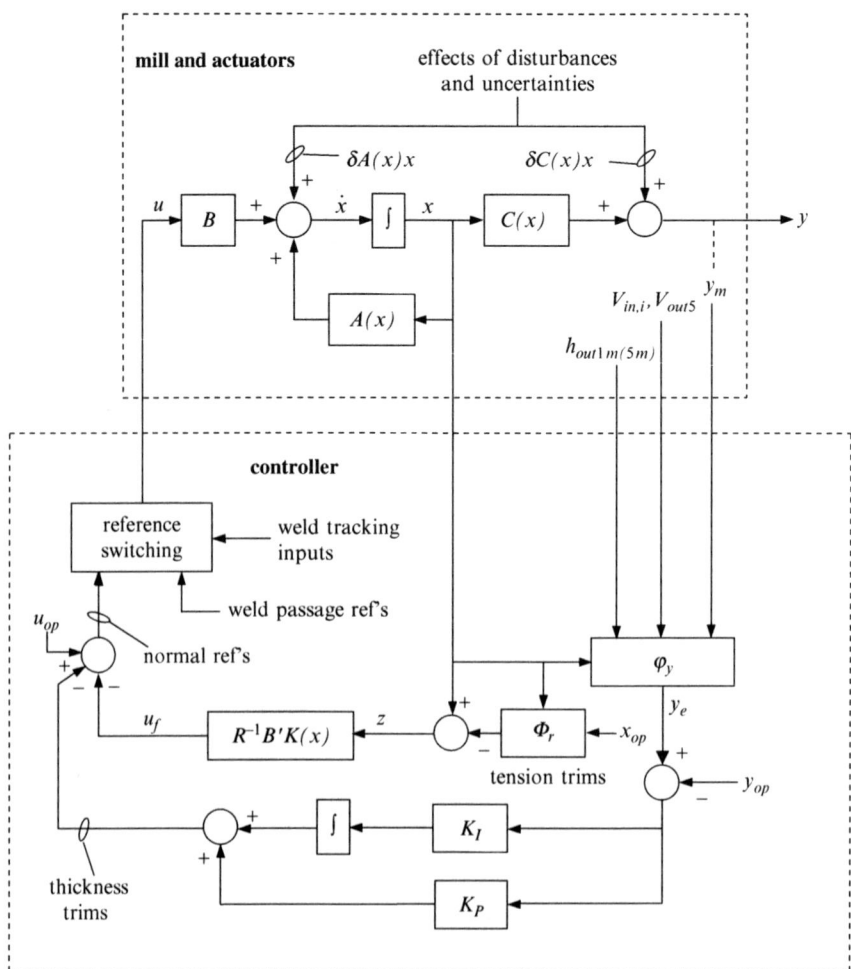

Fig. 5.10 System configuration for continuous tandem cold mill

is based on the basic techniques of continuous control theory, so that the discretized system looks and behaves very much like a continuous controller. Often more than one digital scan is utilized, where the control of the mill is performed in a main scan running fast enough to assure proper control of the process, with any auxiliary scans running asynchronously and usually at slower rates for control of supportive auxiliary functions whose update times can be longer than the time required for updating the main scan.

Of particular interest in the update time of the main scan is the solution of the algebraic Riccati equation (ARE). It is expected that the controller should be sufficiently fast to provide a solution to the ARE during each main scan, or alternatively during an auxiliary scan which runs concurrently with the main controller scan. If such an auxiliary scan is used, it could be asynchronous to the main scan since a solution to the ARE with high accuracy is unnecessary for this application. At present the update time for a typical main scan of an industrial controller for a tandem cold rolling application is about 10–15 ms, which includes the time for the actual controller update, the time for running any necessary auxiliary functions that are precluded from running in slower auxiliary scans, and any overhead required for processing.

To accommodate the timing requirement for solving the ARE, the matrix sign function technique [20] was investigated [19] as a possible viable method for implementing the solution in a controller applied in a real-world setting. The results of simulations performed showed that this method of solving the ARE was quite successful, as the time for solution was about 4 ms using MATLAB\Simulink[4] running on recent software and hardware platforms, which leaves enough time for the remainder of any other tasks needed to be performed during the main scan. Also, the solution time determined during the simulations is consistent with the capabilities of modern industrial controllers. It is expected that as the computational capabilities of industrial controllers continue to improve, the times for solving the ARE online will become even faster, so that issues related to the online solution of the state-dependent Riccati equation will continue to be of negligible concern for this application.

In the actual usage of the controller the trims provide a simple and user-friendly means of adjustment during commissioning, as there is a one-to-one relationship between the trim adjustment and the function being adjusted. Additionally, it is necessary that capability be available to allow the operator to independently adjust the interstand thicknesses and the tensions anywhere in the mill. The one-to-one independent relationship between the trim reference adjustment and the variable being adjusted provides this capability. The results of simulations as described in Section 5.4.2 show that useful ranges of adjustment for this function can be made by using the trims.

5.3.4 Eccentricity Compensation

The implementation of the augmented SDRE controller assumes that there is complete compensation of the eccentricity in the backup rolls and the work rolls of the tandem cold mill. The development of a method to adequately compensate for the eccentricity in each of the rolls under all conditions of operation requires extensive effort that is outside the scope of the work described in this book. However, as numerous papers, patents, and theses describing various methods of eccentricity compensation have been published with successful implementations reported, it is assumed that a method of compensation exists, or could be developed,

---

[4]MATLAB and Simulink are registered trademarks of The MathWorks, Inc., Natick, MA 01760-2098.

to fit within the framework of the SDRE controller and be effective in rejecting the internal disturbances caused by roll eccentricities. To justify this assumption, a technique described in [21] and reported to be implemented successfully in an actual operating multi-stand mill is used as a basis for a conceptual method [18] that compensates for the eccentricity of a backup roll, and has the potential for expansion to compensate for the eccentricities of the other backup rolls and the eccentricities of the work rolls in the mill. The concept is verified by the simulation described in Section 5.4.3.

The roll eccentricity modifies the BISRA relationship (3.38) to add a term to account for the eccentricity as

$$h_{out} = S + S_0 + \frac{F}{M} + e, \tag{5.52}$$

where e is the roll eccentricity. The conceptual method for compensation of e is a form of adaptive noise cancelation that relies on the fact that the backup roll eccentricity is always periodic with a frequency that is proportional to the measured angular velocity of the roll, so that there is correlation between the eccentricity and a sinusoid generated from the measurement of the roll angular velocity.

In general, adaptive noise cancelation is a technique that relies on the correlation between the noise in a noisy signal and the measured noise generated by a separate source. The concept, as described in various texts [e.g., 22, 23], is depicted in Figure 5.11 and is used in the discretized eccentricity compensation (Figure 5.12) which interfaces with the SDRE controller that is discretized for actual usage. In Figure 5.11 n represents the discrete time step, $v_1(n)$ and $v_2(n)$ are correlated noise sources, the signal $y(n)$ is uncorrelated with $v_1(n)$, $v_2(n)$, and $\hat{v}_1(n)$, where $\hat{v}_1(n)$ is the output of the LMS (least mean square) adaptive filter, and it is assumed that $v_2(n)$ contains no components of $y(n)$.

The LMS adaptive filter uses $v_2(n)$ to predict $\hat{v}_1(n)$. The following relationship as depicted in Figure 5.11 is applicable,

$$e_f(n) = \hat{y}(n) = y(n) + (v_1(n) - \hat{v}_1(n)), \tag{5.53}$$

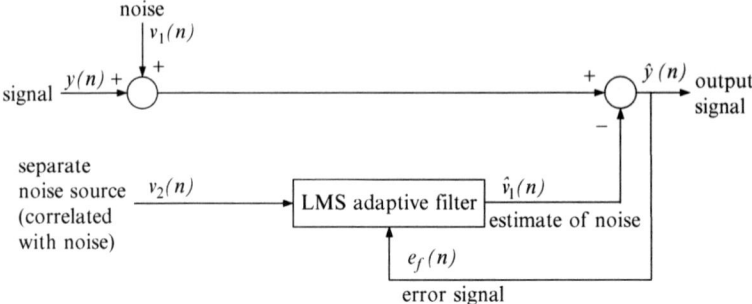

Fig. 5.11 Least mean square adaptive noise cancelation

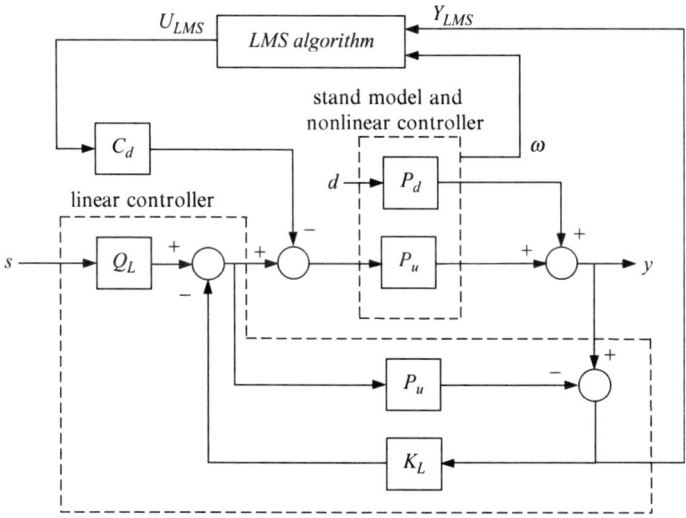

Fig. 5.12 Position controller realization with active eccentricity compensation, based on [21]

where $e_f(n)$ is the error signal applied to the adaptive filter. Squaring and taking expected values gives

$$E[e_f^2(n)] = E[\hat{y}^2(n)] = E[y^2(n)] + E[(v_1(n) - \hat{v}_1(n))^2], \qquad (5.54)$$

where it is noted that the term $2E[y(n)(v_1(n) - \hat{v}_1(n))]$ is zero because $y(n)$ is uncorrelated with $v_1(n)$ and $\hat{v}_1(n)$, and therefore is not shown in (5.54). The LMS adaptive filter will adjust itself to minimize $E[e_f^2(n)]$ and thus minimize $E[(v_1(n) - \hat{v}_1(n))^2]$ while not affecting $y(n)$, which reduces the noise in the output signal $\hat{y}(n)$.

The position actuator controller for a mill stand is taken as an inner control loop for the displacement of the hydraulic spool with a nonlinear inner controller coupled to a linear outer control loop. A factorization approach [21] results in a realization, to which is added the LMS adaptive noise cancelation algorithm, as depicted in Figure 5.12. In Figure 5.12, s is the position actuator reference, y is the position feedback, ω is a signal whose frequency is proportional to the measured angular velocity of the roll, d is the eccentricity disturbance, Y_{LMS} is the input to LMS algorithm, U_{LMS} is the output of the LMS algorithm, P_u is a bounded-input-bounded-output-stable (BIBO-stable) transfer function which describes the input-output linearization obtained by an input transformation in the nonlinear controller, P_d, Q_L, K_L, and C_d are BIBO-stable transfer functions where C_d is determined such that $|C_d(j\omega)P_u(j\omega) - 1|$ is nearly zero in the range of frequencies pertinent to the roll eccentricity. The intent is that the disturbance signal $P_d d$ is canceled by the signal $P_u C_d U_{LMS}$ so that the output y is essentially eccentricity free and the changes in the strip thickness due to the eccentricity are reduced.

Further insight into the use of the LMS algorithm in the eccentricity compensation can be gained by comparing the signals in Figure 5.12 with those in Figure 5.11. In Figure 5.12 the signals Y_{LMS}, $P_u C_d U_{LMS}$, and $P_d d$ correspond to the signals $e_f(n)$, $\hat{v}_1(n)$, and $v_1(n)$, respectively, in Figure 5.11, with $y(n)$ being zero, and a sinusoid generated using the frequency ω corresponding to $v_2(n)$. Thus the estimated eccentricity signal $P_u C_d U_{LMS}$, corresponding to the estimated noise signal $\hat{v}_1(n)$, is subtracted from the eccentricity signal $P_d d$, corresponding to the noise signal $v_1(n)$, to reduce the eccentricity component in the output y. In addition, it is assumed that the eccentricity noise in the signal representing roll force, and any remaining eccentricity noise in the signal representing actuator position, are reduced by adaptive filtering techniques [e.g., 24], so that the signals representing roll force and actuator position both have negligible eccentricity components. The effects on eccentricity compensation caused by changes in roll diameter due to heating and mechanical wear, and by harmonics in the eccentricity waveform, are addressed in the simulations (Section 5.4.3). This method can be extended to estimate the eccentricity of the other backup roll and the eccentricities of the work rolls by using the signal ω to generate sinusoids of appropriate frequencies.

5.3.5   Brief Evaluation of Other Advanced Methods

Chapter 4 presented two advanced methods for control of the tandem cold rolling process. This section provides a brief summary of the strengths and weaknesses of each of these. In addition some material describing other advanced methods for possible control of this process is included. Of course not every advanced method can be considered. Material briefly describing certain other additional advanced techniques which could be evaluated for the control of this process can be found in [13]. The methods addressed herein are some of those which have been applied very successfully to the control of many other industrial processes, and therefore might be seen as offering the potential for viable control of this process. The summaries of each method are brief and are intended to give an overall picture of some of the main features of each of the methods being addressed, which can be evaluated in light of the characteristics of the augmented SDRE method as previously presented. Criteria that can be considered for evaluation of an advanced method of control for the tandem cold rolling process are the following:

- The control method should result in an improvement in performance over the better performing conventional methods during the various regimes of mill operation, considering realistic disturbances and uncertainties.
- It should be able to handle rapid changes in product, especially in the application to continuous tandem cold rolling where product changes occur rapidly on the fly.
- It also should provide a controller structure that is user-friendly to design and commissioning personnel, most of whom have limited backgrounds in advanced

control theory. This is especially important during commissioning as a reduction in the startup time is essential to realizing a timely entry into profitable production.

- It is preferred that it use a nonlinear control strategy to reduce the computation required for linearization coefficients that change with operating conditions and product.
- It should conform to other significant criteria that are specific to the method being considered.

5.3.5.1 The H^{∞} Loop Shaping Approach

The H^{∞} loop shaping approach is described in Section 4.3. In Section 4.3.5 comments on this approach are included. As noted therein, simulation based on the use of the H^{∞} loop shaping approach has realized an improvement over the simulation of a conventional method for operation at a certain operating point and during speed change from run to thread and the reverse. However, this capability was demonstrated with only one product. The redesign for a variety of different products would require an online update for the selection of suitable weights for each product or group of products. Selection of these weights could require considerable additional design effort to accommodate continuous tandem cold rolling where product changes occur quite rapidly, which would require rapid changes in the controller. This puts a greater burden on the control designers who often have a limited background in advanced control theory, particularly in the selection of weights, as the choice of suitable weightings is not a simple task and requires some experience and intuition in this area. Moreover, accommodating the change in speed could require the gain scheduling of several controllers which adds to the design complexity. As noted in Section 4.3.5 these complexities, plus the inherent complexity of a MIMO controller that includes dynamics, poses difficulties in tuning, especially to most commissioning personnel who have very little or no background in advanced control theory and are accustomed to simpler configurations, such as a one-to-one relationship between a tuning adjustment and an associated process variable. In addition, a linearized model is required.

5.3.5.2 An Observer-based Method for MIMO Control

This method is described in Section 4.4, with comments in Section 4.4.5. An improvement in performance over a conventional method has been realized in an actual application on a two-stand mill. The major improvement noted in this method is the reduction in deviations in the mill output thickness during speed change, with only slight improvement in the reduction of deviations during steady speed. The ability to handle rapid changes in product was not addressed in the description but could require the development of additional algorithms for establishing controller settings for the next product prior to the transition between products, and to implement the new settings during the transition. The controller complexity could present difficulties in tuning at commissioning. Additionally, while improved performance during speed change has been realized, further improvement during speed change and steady speed might be achievable if the time delays were included in the actual model and the disturbances and uncertainties were treated separately. The two-stand configuration uses a thickness measurement between the stands which is inappropriate for most tandem mills with additional stands, as a thickness measurement usually is not used between each pair of adjacent stands due to space limitations and the high cost of the additional measurement systems. A linearized model is required.

5.3.5.3 Model Predictive Control

Model predictive control techniques have been developed for both linear and nonlinear applications, with control for the linear applications being the most highly developed. The basic idea of model predictive control is that the control inputs are determined using an optimization criterion that is developed based on a prediction horizon. The system model is used to predict the effects of the future inputs on the outputs and on the states of the system. More specifically and assuming a discrete implementation, an objective function is formed using the process model. A move of the controller is then computed for each of the discrete steps in a prediction horizon so as to minimize the objective function, such that future values of the outputs and the states are estimated while including the model as a constraint in the form of the state equation. However, only the inputs computed for the first time step are actually implemented. The prediction horizon is then moved forward one time step and the procedure is repeated using the measurements that resulted from the first movement of the controller.

Model predictive control has seen many successful applications in the chemical process industries and various other applications. However, while this method has several features that could be useful in the control of tandem cold rolling, and some methods for reducing the online computation time of this technique have been proposed, a controller update time on the order of 10–15 ms presently appears to be beyond its capability to assure the desired performance of a large nonlinear fast system such as the tandem cold rolling process which has significant time delays, uncertainties, and disturbances. In addition, the complexity of the controller would result in difficulties in tuning, especially if a nonlinear approach was used. More detailed information on model predictive control is found in [25, 26].

5.3.5.4 Sliding Mode Control

Sliding mode control is characterized by a group of feedback control laws and a switching function. The input to the switching function is based on some measure of the current system behavior. Based on this input, the switching function produces an output that represents the feedback control law that should be used at that particular time, *i.e.* the control law is deliberately changed in accordance with some predefined rules that depend on the state of the system. Initially a manifold of a dimension lower than that of the original system is determined which can be stabilized and made positive invariant by a suitable control input. Then a control law is chosen that forces the system states that are not initially on the manifold to go to the manifold in a finite time. The control input is switched between an upper and lower bound, with the switching logic being determined depending on the distance from the manifold. Since the manifold is invariant, once the trajectory gets to the manifold it remains there, and there is no longer sensitivity to uncertainties in the inputs. The system's behavior when on the manifold, *i.e.* the sliding surface, is denoted as the sliding mode. However, because of process characteristics such as time delays and hysteresis effects, the system's states never remain on the manifold but repeatedly cross it to produce an effect denoted as chattering. In response to this, the control action is modified to force the states to remain within a boundary layer around the manifold. The addition of the boundary layer to prevent chattering causes the robustness to uncertainties to be reduced somewhat depending on the thickness of the boundary layer.

Sliding mode control has been applied very successfully to many industrial applications which require robustness to disturbances and modeling uncertainties. In general most of these applications were much smaller than the tandem cold rolling process with fewer system parameters, and therefore required simpler controller structures. The selection of the boundary layer is important to preclude chattering which in the case of tandem cold rolling is unacceptable. In larger systems where the system uncertainties and disturbances must be estimated, the determination of a boundary layer which is effective in attaining good robustness and yet absolutely precludes chattering is difficult considering that most of the disturbances and uncertainties cannot be measured, change significantly with operating conditions, and interact with each other, which makes this method of control less desirable for tandem cold rolling. Additional material on the theory and application of sliding mode control can be found in [27, 28].

5.4   Simulations

In this section the results of closed-loop simulations are presented to provide an indication of the performance of the augmented SDRE controller when coupled to the model of the tandem cold rolling process. Results of other simulations related to the estimation of output thickness, eccentricity compensation, and transitions between products during continuous tandem rolling also are included. Reports of simulations and their results are given in [29–32].

5.4.1 Verification of Performance

Simulations were done to verify the performance of the controller during the various regimes of mill operation and with the disturbances and uncertainties previously described. To give some feel for the effectiveness of the trims in reducing excursions in thicknesses and tensions, initial simulations were performed with no disturbances or uncertainties, with the mill speed being changed from run to thread and then the reverse, with the trims configured and with tuning done to reduce the excursions in the thicknesses and interstand tensions. The change in speed was effected by using a mill master speed reference to change the work roll speeds simultaneously in accordance with values calculated based on the product reductions, the forward slips as estimated using the model developed in Chapter 2, and the mill data given in Chapter 2. Table 5.8 lists the forward slips, output thicknesses, and peripheral speeds of each roll at a run speed of 1,220 m/min at the mill exit, which corresponds to a 100% of the mill master speed reference. The speed reference applied to each work roll speed actuator was proportional to the mill master speed reference.

The master speed reference was shaped to reduce excursions in tensions at the start and at the finish of a speed change. The times for speed changes are consistent with typical rates for a stand-alone mill. Figures 5.13 and 5.14 show the shape and

Table 5.8 Mill rolling data at 1,220 m/min

Stand	Forward slips, output thicknesses, and roll speeds at 1,220 m/min				
	1	2	3	4	5
Forward slip	0.045	0.023	0.020	0.022	0.003
Output thickness, mm	2.95	2.44	2.01	1.68	1.58
Peripheral roll speed, m/min	625.0	772.4	940.1	1123.0	1216.6

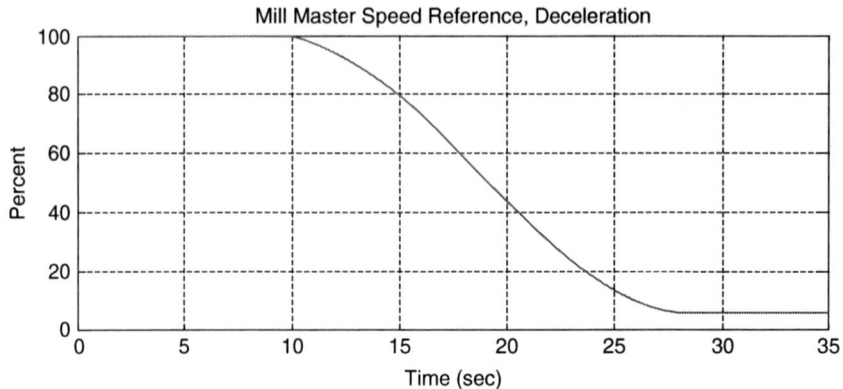

Fig. 5.13 Mill master speed reference during deceleration

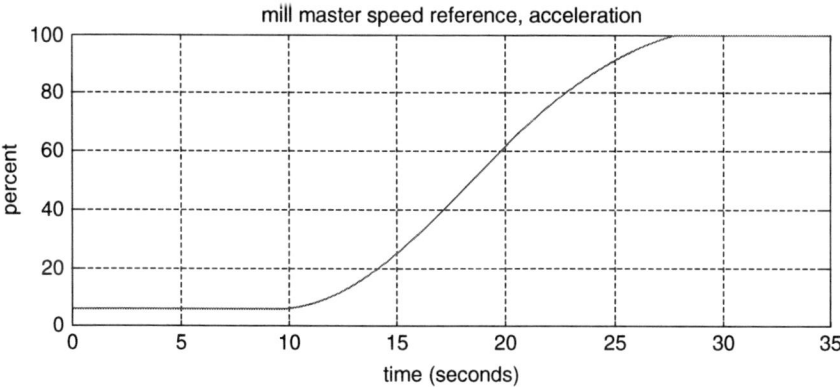

Fig. 5.14 Mill master speed reference during acceleration

Table 5.9 Deviations in thicknesses and tensions, without disturbances or uncertainties

Variable	Magnitude of maximum percent deviation of variable from operating point value			
	Run speed	Thread speed	Run to thread speed	Thread to run speed
h_{out1}	0%	0	<0.001	<0.001
h_{out2}	0	0	<0.001	<0.001
h_{out3}	0	0	<0.001	<0.001
h_{out4}	0	0	<0.001	<0.001
h_{out5}	0	0	<0.001	<0.001
σ_{12}	0.03	0.02	0.03	0.03
σ_{23}	0.04	0.00	0.05	0.05
σ_{34}	0.01	0.01	0.02	0.02
σ_{45}	0.10	0.05	0.10	0.10

the percent change in the master speed reference during deceleration from run to thread, and from thread to run.

The operating point was as noted in Table 2.2 Production Schedule 1, with the mill and strip parameters taken as in Table 2.3. Operating Point 1 of Table 2.4 was also simulated with insignificant difference noted in the results between Operating Point 1 and Production Schedule 1. The settings of the Q and R matrices and the gain settings of the trims were made based on physical intuition and a few trials to arrive at settings which reduced undesirable excursions in thicknesses and tensions. In all cases the eccentricity compensation and the active filtering function were assumed to be fully effective so that the effects of eccentricity were taken to be insignificant. Table 5.9 summarizes the results, where run speed is 1,220 m/min and thread speed is 61 m/min.

To get a feel of how the excursions in thickness and tension depend on the disturbances, the disturbances in the incoming thickness and hardness without uncertainties then were applied and the simulations repeated. The results are summarized in Table 5.10 and in Figures 5.15–5.22.

Table 5.10 Deviations in thicknesses and tensions with disturbances, without uncertainties

Variable	Magnitude of maximum percent deviation of variable from operating point value			
	Run speed	Thread speed	Run to thread speed	Thread to run speed
h_{out1}	0.022%	0.005	0.021	0.025
h_{out2}	0.012	<0.001	0.011	0.011
h_{out3}	0.014	<0.001	0.012	0.013
h_{out4}	0.015	<0.001	0.014	0.013
h_{out5}	0.011	<0.001	0.010	0.010
σ_{12}	0.11	0.02	0.10	0.10
σ_{23}	0.05	0.00	0.05	0.05
σ_{34}	0.04	0.01	0.04	0.04
σ_{45}	0.14	0.05	0.10	0.10

The uncertainties in modeling and measurement then were applied concurrently with the disturbances. The uncertainties were combined with each other and with the disturbances in a manner to approximate the more severe conditions which include significant excursions in the mill exit thickness plus significant excursions in the exit thicknesses of the intermediate stands and in the interstand tensions. The results are summarized in Table 5.11.

The previous simulations were repeated for Production Schedule 2 and Production Schedule 3 as listed in Table 2.2. For the case of Production Schedule 2, changes in the settings of the Q and R matrices and the trims were unnecessary, and good performance was realized with the settings remaining as in the simulations for Production Schedule 1. In the case of Production Schedule 3 the settings of Q and R remained unchanged. The settings of $K_{I,i}$, $K_{P,i}$, K_{g_int1}, and K_{g_int5} were changed slightly to improve the performance. The results are summarized in Tables 5.12 and 5.13.

5.4.2 Verification of the Output Thickness Estimate and the Independence of Operator Adjustments

5.4.2.1 Verification of the Output Thickness Estimate

The relationship (5.38) for output thickness uses the strip speed $V_{in,i+1}$ at the input of stand $i + 1$ in place of the strip speed $V_{out,i}$ at the output of stand i as in (5.37), i.e.

$$h_{out,i} = \frac{h_{in,i}V_{in,i}}{V_{in,i+1}}k_{i,e}, \quad (i = 2, 3, 4), \tag{5.55}$$

is taken as equivalent to

$$h_{out,i} = \frac{h_{in,i}V_{in,i}}{V_{out,i}}k_{i,e}, \quad (i = 2, 3, 4), \tag{5.56}$$

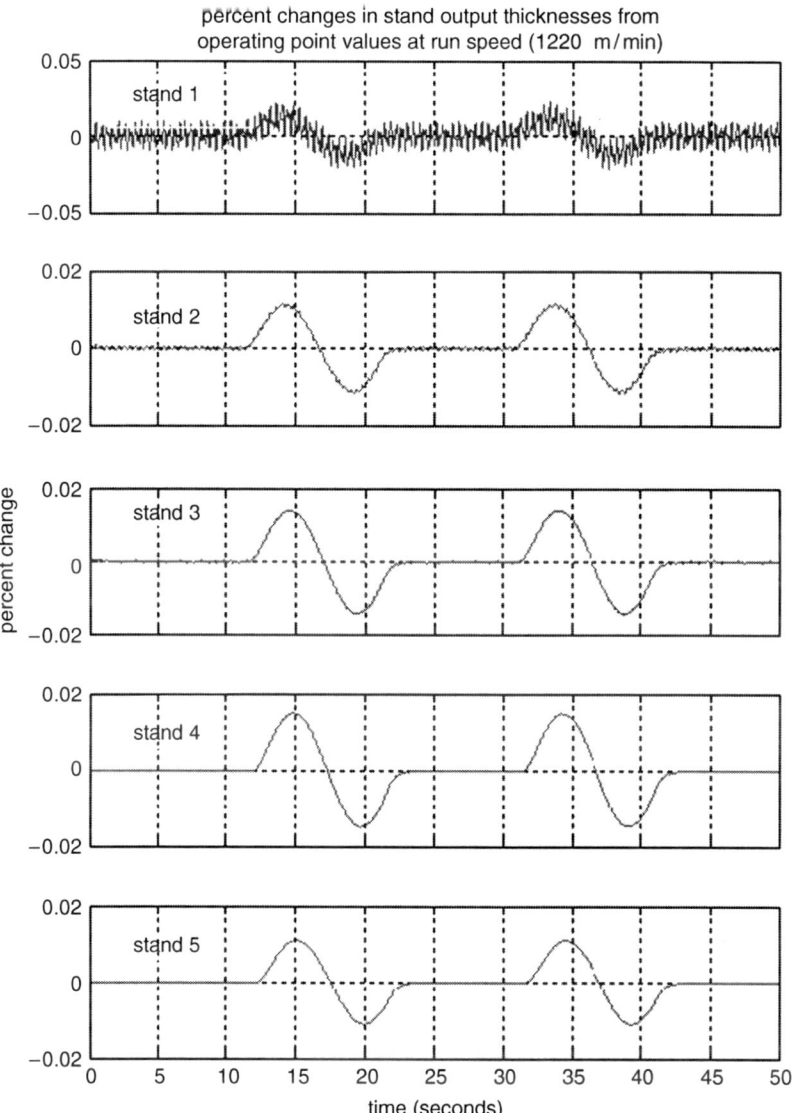

Fig. 5.15 Stand output thickness responses to external disturbances at run speed

where (5.55) and (5.56) are copies of (5.38) and (5.37) respectively. The relationship (5.38) is verified by repeating the simulations of Section 5.4.1 with disturbances and uncertainties applied using (5.55), and then assuming that measurements of $V_{out,i}$ exist, again repeating these simulations with disturbances and uncertainties applied using (5.56). The magnitudes of the maximum deviations in the responses with (5.55) from the responses with (5.56) give an indication of the validity of

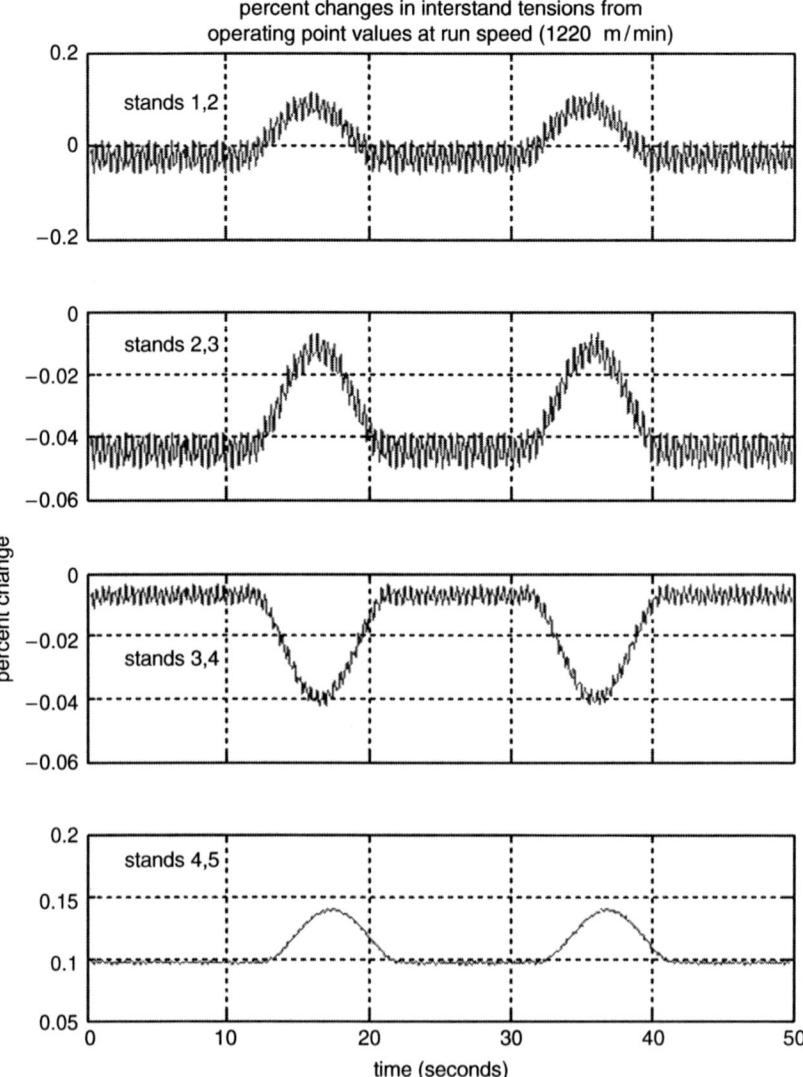

Fig. 5.16 Interstand tension responses to external disturbances at run speed

taking (5.55) in place of (5.56). The results of these simulations are that, using (5.55) in place of (5.56), the largest magnitude of the maximum deviation in stand exit thickness is negligible (*i.e.* less than 0.001%) and the largest magnitude of the maximum deviation in interstand tension also is negligible which provides the justification for using (5.38) in place of (5.37).

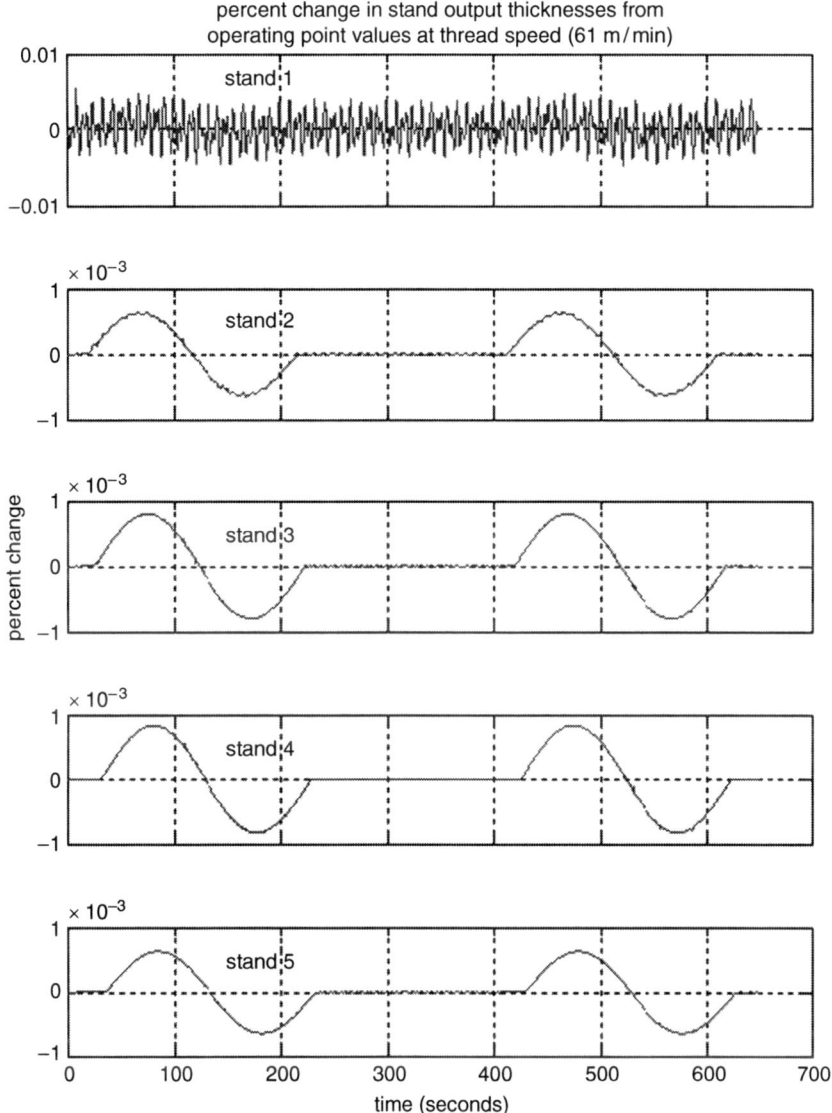

Fig. 5.17 Stand output thickness responses to external disturbances at thread speed

5.4.2.2 Verification of the Independence of the Operator Adjustments

It is important in the operation of the mill to provide the operator the capability to independently adjust a strip thickness or a strip tension anywhere in the mill. This capability was confirmed by simulation which showed that an adjustment of 2% in the output thickness of any stand resulted in a negligible effect on the steady-state

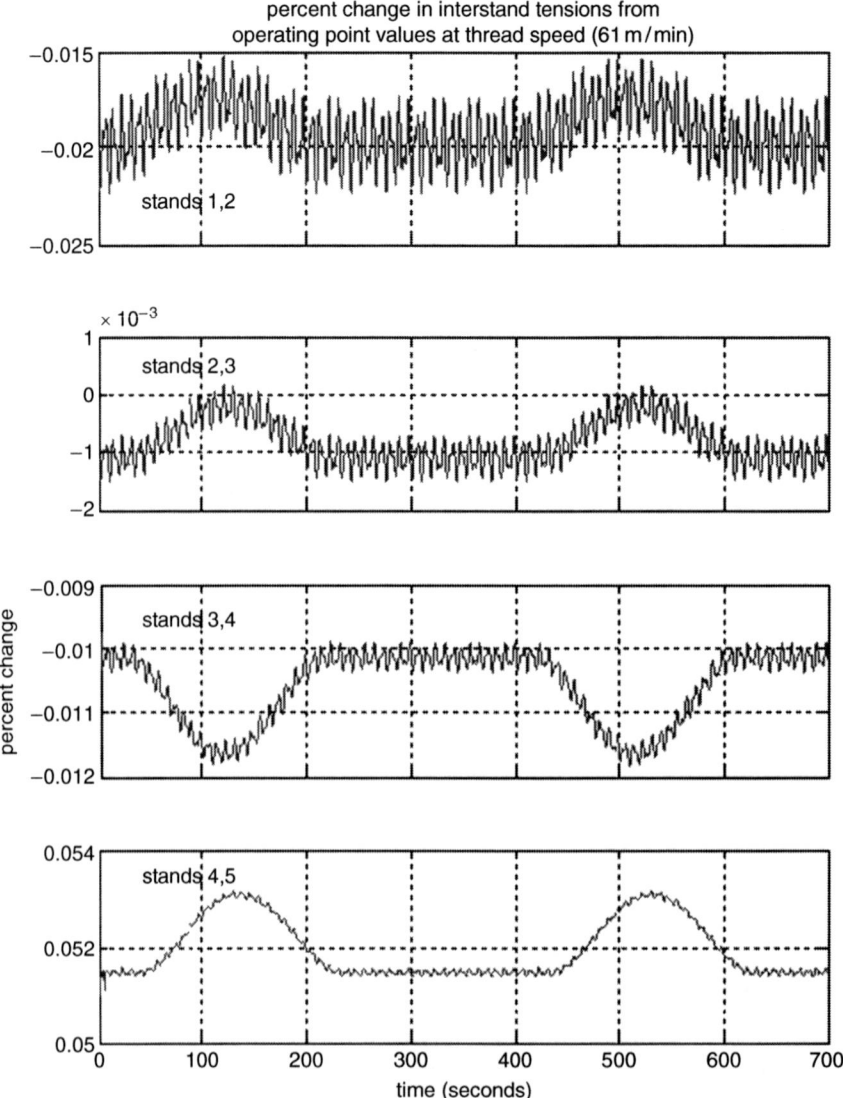

Fig. 5.18 Interstand tension responses to external disturbances at thread speed

thicknesses and tensions which were not being adjusted. A similar effect was
verified for adjustment of the tensions, where a 5% adjustment was confirmed.

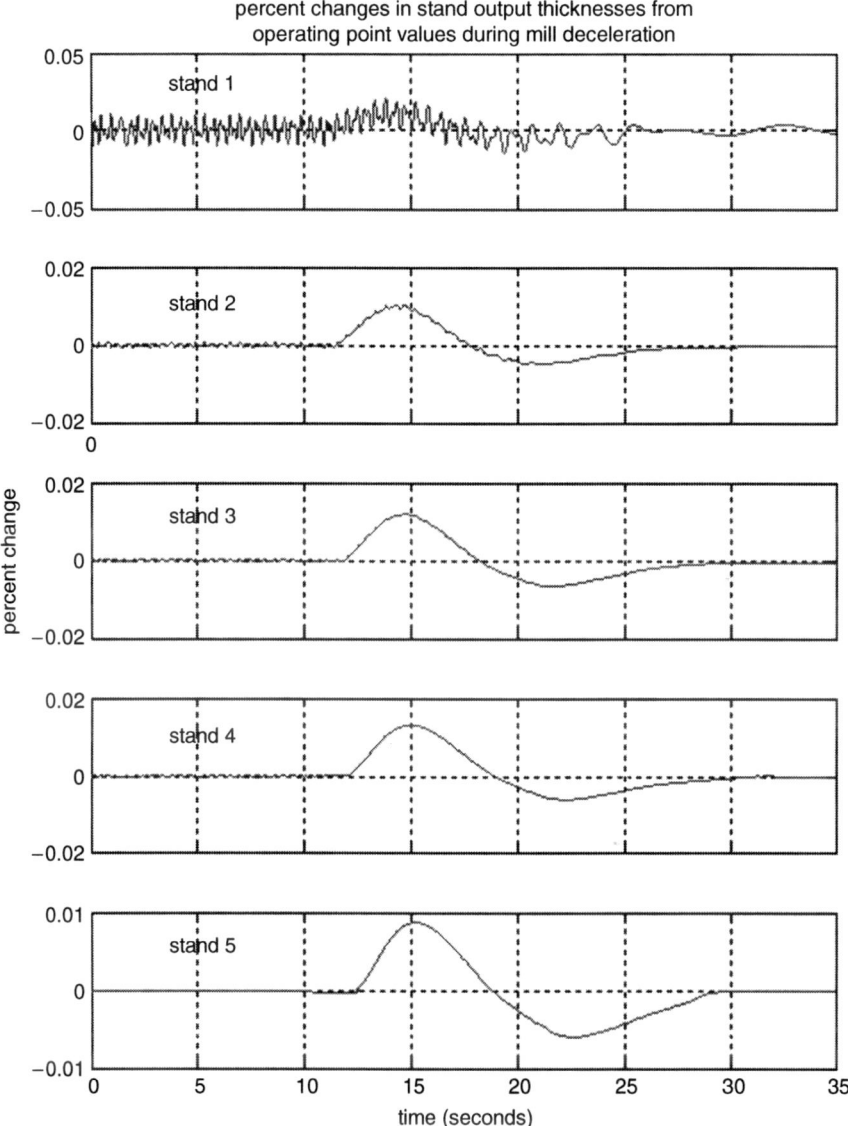

Fig. 5.19 Stand output thickness responses to external disturbances during decel from run speed to thread speed

5.4.3   Verification of the Eccentricity Compensation

The following assumptions were used to verify performance of the conceptual eccentricity compensation technique described in Section 5.3.4:

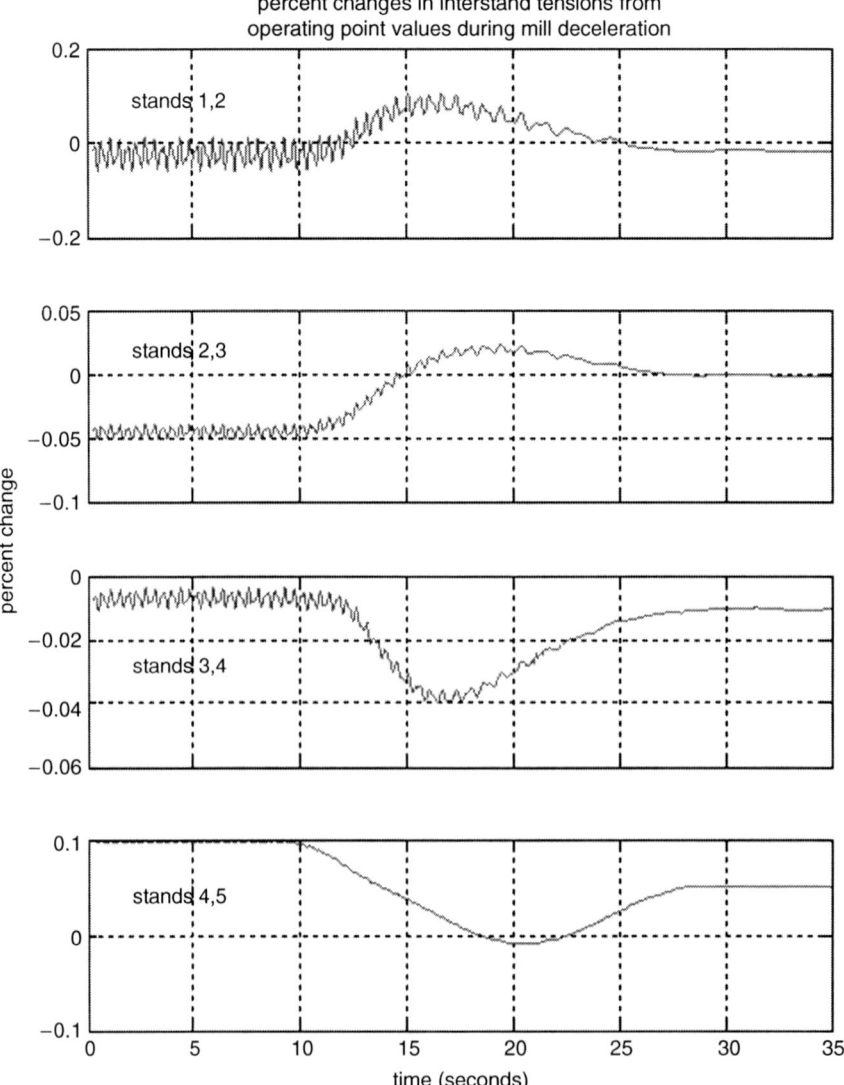

Fig. 5.20 Interstand tension responses to external disturbances during decel from run speed to thread speed

- The eccentricity is only in the backup rolls, which have identical eccentricity. The work rolls are eccentricity free.
- The diameter of the backup rolls is 1,346 mm which can change about 2.54 mm (0.2%) due to the effects of mechanical wear [33]. For the simulation a change of 0.5% in the diameter is assumed for other effects and conservatism.

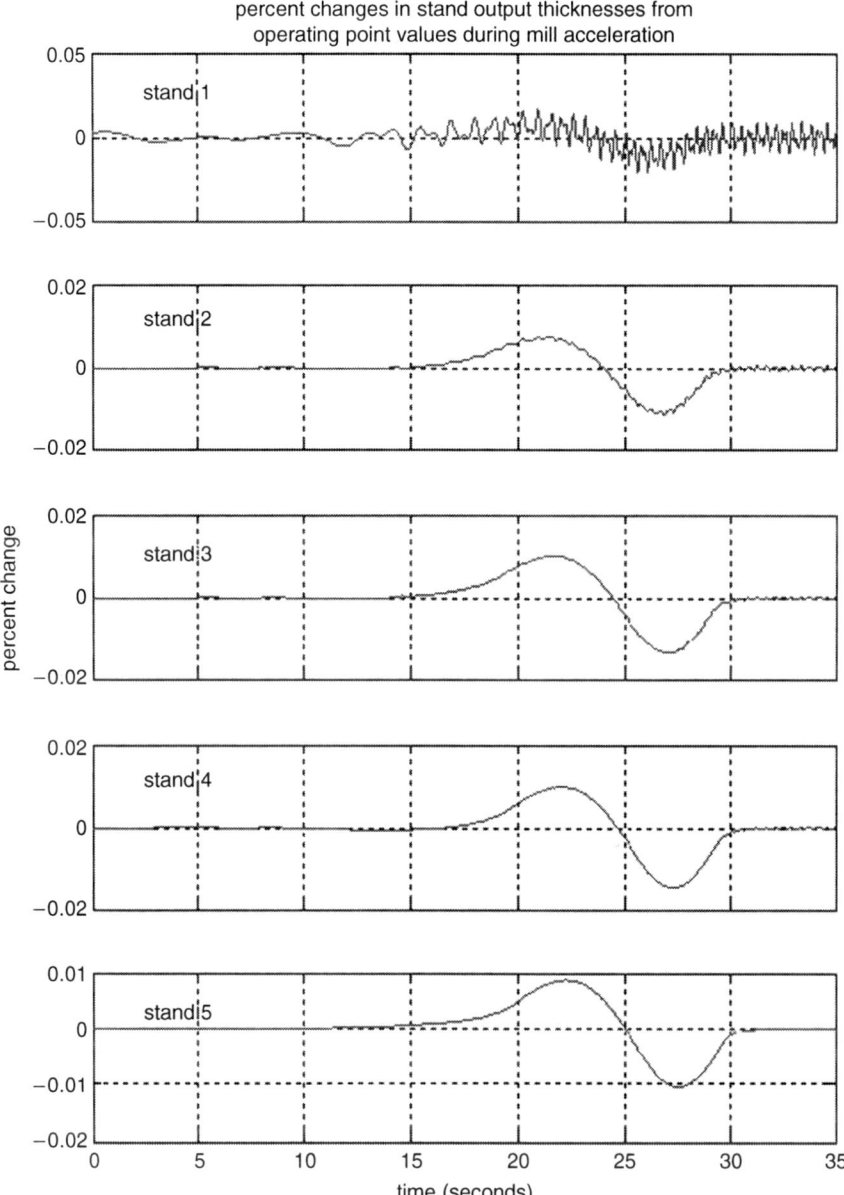

Fig. 5.21 Stand output thickness responses to external disturbances during accel from thread speed to run speed

- The eccentricity is a sinusoid plus a third harmonic. The fundamental is taken to have a period corresponding to one revolution of the backup rolls with a peak of 0.03 mm, which is about 2% [34] of the operating point mill exit thickness of

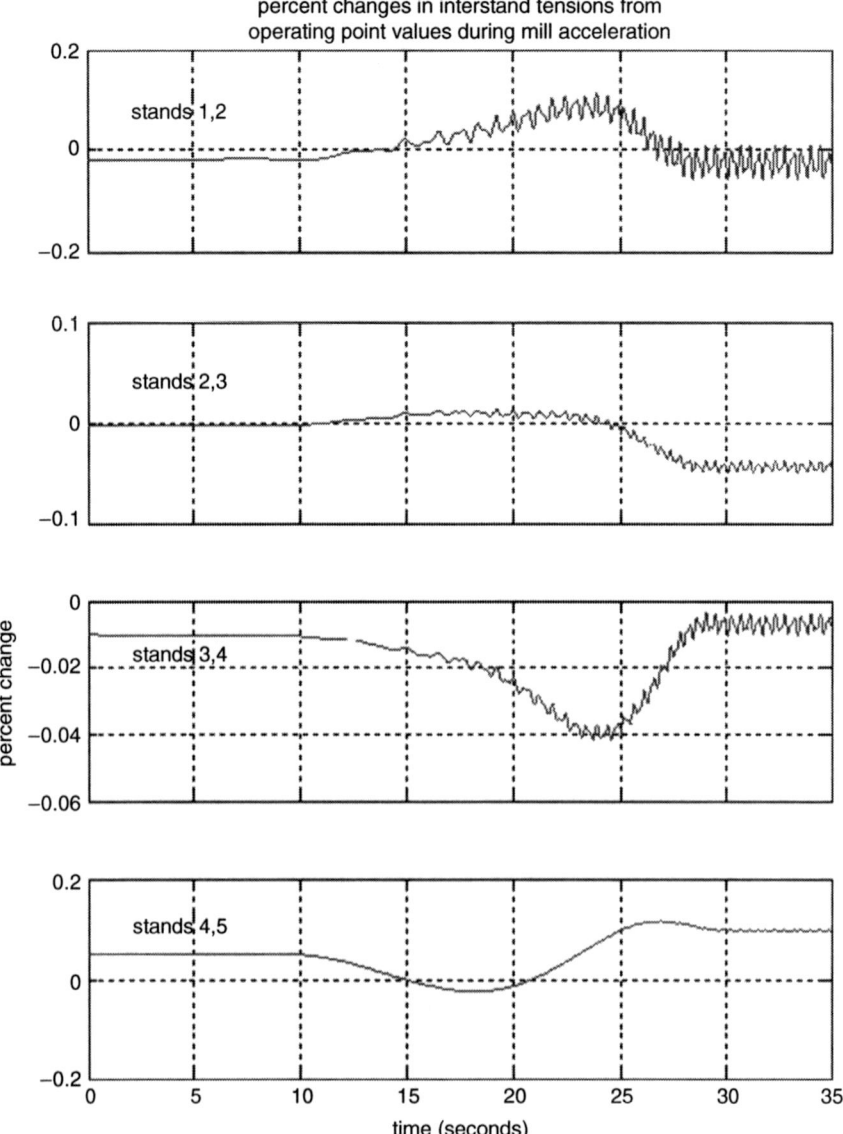

Fig. 5.22 Interstand tension responses to external disturbances during accel from thread speed to run speed

1.575 mm. The peak of the third harmonic is taken to be 3% of the peak of the fundamental. The same eccentricity is assumed for each mill stand.

- The symbols used in this section are those used in Section 5.3.4, unless otherwise noted.

Table 5.11 Deviations in thicknesses and tensions with disturbances and uncertainties applied, Production Schedule 1

Variable	Magnitude of maximum percent deviation of variable from operating point value			
	Run speed	Thread speed	Run to thread speed	Thread to run speed
h_{out1}	0.160%	0.020	0.141	0.100
h_{out2}	0.073	0.051	0.062	0.072
h_{out3}	0.079	0.051	0.075	0.068
h_{out4}	0.081	0.052	0.063	0.079
h_{out5}	0.072	0.058	0.074	0.071
σ_{12}	0.26	0.04	0.20	0.45
σ_{23}	0.12	0.02	0.11	0.21
σ_{34}	0.21	0.02	0.20	0.41
σ_{45}	0.09	0.02	0.04	0.15

Table 5.12 Deviations in thicknesses and tensions with disturbances and uncertainties applied, Production Schedule 2

Variable	Magnitude of maximum percent deviation of variable from operating point value			
	Run speed	Thread speed	Run to thread speed	Thread to run speed
h_{out1}	0.221%	0.025	0.210	0.121
h_{out2}	0.084	0.052	0.065	0.082
h_{out3}	0.092	0.052	0.087	0.077
h_{out4}	0.110	0.053	0.071	0.097
h_{out5}	0.082	0.051	0.079	0.071
σ_{12}	0.23	0.02	0.20	0.22
σ_{23}	0.15	0.01	0.13	0.14
σ_{34}	0.15	0.00	0.15	0.16
σ_{45}	0.44	0.04	0.45	0.43

Table 5.13 Deviations in thicknesses and tensions with disturbances and uncertainties applied, Production Schedule 3

Variable	Magnitude of maximum percent deviation of variable from operating point value			
	Run speed	Thread speed	Run to thread speed	Thread to run speed
h_{out1}	0.275%	0.045	0.250	0.142
h_{out2}	0.078	0.052	0.062	0.078
h_{out3}	0.080	0.052	0.077	0.071
h_{out4}	0.089	0.053	0.065	0.090
h_{out5}	0.078	0.051	0.071	0.070
σ_{12}	0.18	0.02	0.18	0.20
σ_{23}	0.35	0.02	0.35	0.35
σ_{34}	0.14	0.00	0.12	0.14
σ_{45}	0.80	0.07	0.80	0.78

The method of eccentricity compensation uses the LMS (least mean square) adaptive filtering technique which is well known and described in various texts [e.g., 22, 23] on statistical digital signal processing and adaptive filtering. For the simulations, a modified normalized LMS algorithm [22] is used to update the filter coefficients. In this algorithm (5.57), the gradient step size is normalized with respect to the norm of the input vector to reduce the effects of gradient noise amplification, and a small positive constant is added to the denominator of the correction applied to the filter coefficient to prevent similar effects if the norm of the input vector becomes too small. The expression for the algorithm is

$$w_{n+1} = w_n + \frac{\beta X_2^*(n)}{\varepsilon + \|X_2(n)\|^2} e_f(n), \tag{5.57}$$

where n represents the discrete time step, w_n is the filter coefficient, β is the normalized gradient step size, ε is a small positive constant, $X_2(n)$ is the filter input vector, and $e_f(n)$ is the filter error (scalar). $X_2^*(n)$ is the complex conjugate of $X_2(n)$, which is equal to $X_2(n)$ in this case.

The LMS filter used in simulations was order 12 with a β of 0.5 and a sampling rate of at least 50 samples per period of the sinusoid assumed for the eccentricity. Initially (Case 1) the eccentricity $v_1(n)$ was assumed to be equal to the sinusoid $v_2(n)$ generated using the speed of the backup roll, as inferred from the measured speed of the work roll, with no harmonic. For this case, the resulting eccentricity after compensation was negligible (less than 0.003%), following filter learning which occurred in less than two revolutions of the backup roll. The learning curve of the filter for this case is plotted in Figure 5.23. The following variations from Case 1 were then individually simulated:

- Case 2: The magnitude of $v_1(n)$ is twice the magnitude of the magnitude of $v_2(n)$.
- Case 3: The magnitude of $v_1(n)$ is one-half the magnitude of the magnitude of $v_2(n)$.

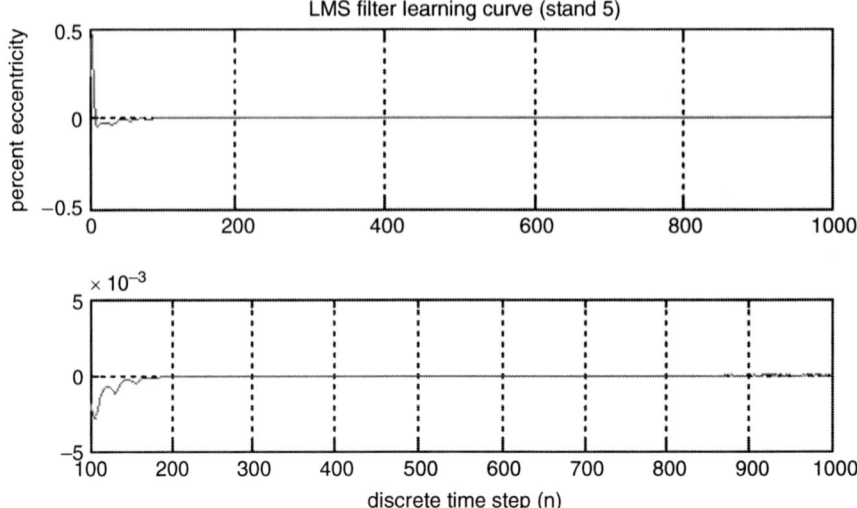

Fig. 5.23 Stand 5 percent eccentricity after compensation, Case 1

- Case 4: The frequency of $v_1(n)$ is 0.5% greater than the frequency of $v_2(n)$.
- Case 5: The frequency of $v_1(n)$ is 0.5% less than the frequency of $v_2(n)$.
- Cases 6 through 8: $v_2(n)$ is shifted by N_i ($i = 1,2,3$) time steps, where N_i is $N/4$ rounded to the nearest integer, with N equal to the number of time steps for one period of the sinusoid.
- Case 9: The third harmonic is added to the fundamental.
- Case 10: The results of the above were combined such that approximately the greatest deviation in eccentricity after compensation was realized.

The results are presented in Table 5.14. As can be seen from these results, the maximum eccentricity remaining after compensation is about 0.05% which supports the assumption that a workable eccentricity compensation technique compatible with the SDRE controller can be developed.

5.4.4 Simulations Relating to the Continuous Process

The areas of interest during weld passage, which marks the transition from the present strip to the next strip in the continuous tandem cold rolling process, are the capability of the controller to reduce the out of tolerance length of strip, reduce excursions in the specific roll force and tension, and maintain mass flow in the mill during the transition, using the first method as described in Chapter 3. The simulations and the results described in this section are based on work reported in [31] wherein the first method is used. Since the second method described in Chapter 3 has some similarities to the first method, but has less severe requirements as the weld passes with the roll gap open, it is considered that a portion of the simulations done for the first method also have application to the second method.

5.4.4.1 Identification of a Preferred Path for the Roll Gap Actuator Position

The reduction in the lengths of strip in the vicinity of the weld that are out of tolerance and in the excursions in the specific roll force requires the identification of a preferred

Table 5.14 Magnitude of maximum eccentricity after filter learning, stand 5

Case	Magnitude of max eccentricity
1	0.003%
2	0.006
3	0.002
4	0.01
5	0.01
6	0.02
7	0.003
8	0.02
9	0.05
10	0.05

path for the movement of the roll gap position actuator during the transition. Accordingly, simulations were performed at stand 1 to identify such a preferred path. Five different paths were selected for evaluation. The simulations were performed with no disturbances or uncertainties, with the operating points for the two strips as shown in Table 5.15 with a thickness transition of 20%. During the simulations the mill was at a weld transition speed of about 122 m/min (10% of run speed), and the position actuator was moved at its maximum speed (1.5 mm/s) to reduce the out of tolerance length. The results are summarized in Table 5.16, and some typical plots are shown in Figures 5.24–5.26. Path 2 was selected as the preferred path. This is based on reducing the out of tolerance length, the peak specific roll force, and the maximum change in the specific roll force. For each of the five paths it is noted that the rapid jump in specific roll force as the weld goes through the roll bite is about the same (about 1.95 kN/mm width). Based on the results obtained for stand 1, Path 2 also was selected as the preferred path for stands 2 through 5.

5.4.4.2 Verification of Performance

Simulations were performed using the preferred path, with increases in the thickness of 5%, 10%, and 20% from the present strip to the next strip. The position actuator movement was initiated such that the weld passes through the roll bite when the actuator was approximately at one-half of its travel. This allows for some margin

Table 5.15 Operating points, present strip and next strip

Variable	Present strip	Next strip, with thickness transition of		
		5%	10%	20%
h_{in1}	3.56 mm	3.73	3.91	4.32
h_{out1}	2.95	3.10	3.23	3.56
h_{out2}	2.44	2.56	2.67	2.79
h_{out3}	2.01	2.11	2.21	2.29
h_{out4}	1.67	1.75	1.78	1.83
h_{out5}	1.58	1.65	1.68	1.73
σ_{10}	0.024 kN/mm^2	0.024	0.024	0.024
σ_{12}	0.086	0.086	0.086	0.085
σ_{23}	0.088	0.088	0.087	0.086
σ_{34}	0.089	0.089	0.089	0.088
σ_{45}	0.092	0.092	0.092	0.089
σ_{50}	0.028	0.028	0.028	0.028

Table 5.16 Simulation results, paths 1 through 5

Path number	Out-of-tolerance length	Max change in specific roll force	Peak specific roll force
1	0.21 m	2.1 kN/m	11.6
2	0.22	2.2	10.7
3	0.66	3.1	10.6
4	1.06	4.1	10.6
5	1.63	5.2	10.5

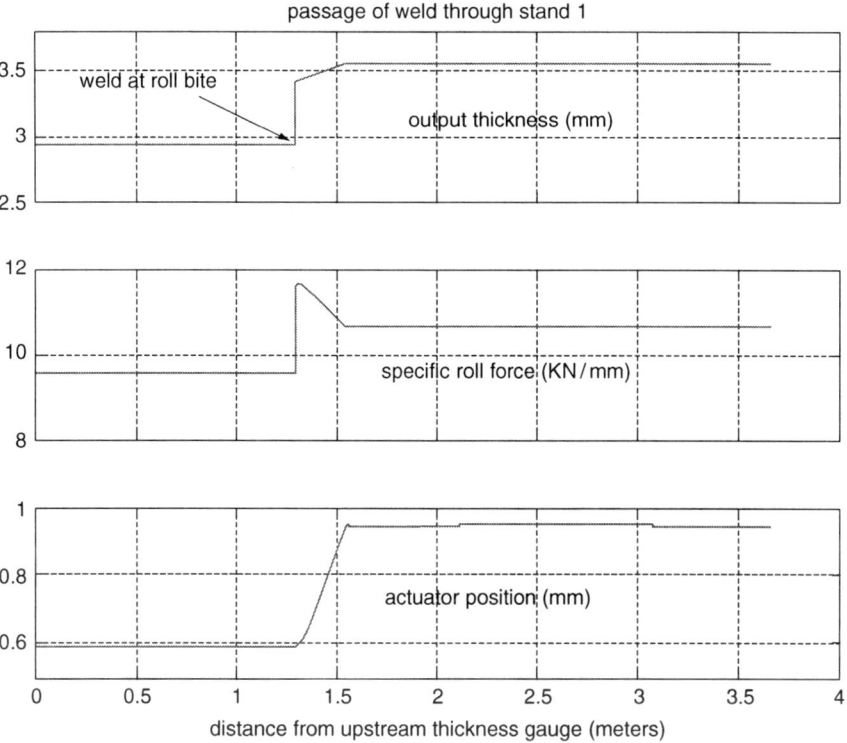

Fig. 5.24 Simulation results for path 1

around the point where the weld passes through the roll bite. The position actuator for stand 1 was moved at its typical maximum speed during the transition. The length of strip passing during the transition at stand 1 was not decreased as the weld moved through the downstream stands. This retained the margin around the half-travel point and could require that the associated position actuators be moved at less than their maximum speeds. If the desired length could not be achieved with the associated position actuator at its maximum speed, then the length of strip passing during a transition at a downstream stand would be increased. Filtering techniques were applied in every case during switching of the controller modes such that the position and speed references would be changed slowly enough to be within the capabilities of the position and speed actuators and their controllers, and to preclude excessive excursions in the associated variables. The simulation results are presented in Tables 5.17–5.19 and typical responses are shown in Figures 5.27 and 5.28.

5.4.4.3 Effects of Disturbances and Uncertainties

During the transition to the next strip a very rapid change in the hardness of the strip which occurs nearly concurrently with the change in the thickness is treated as

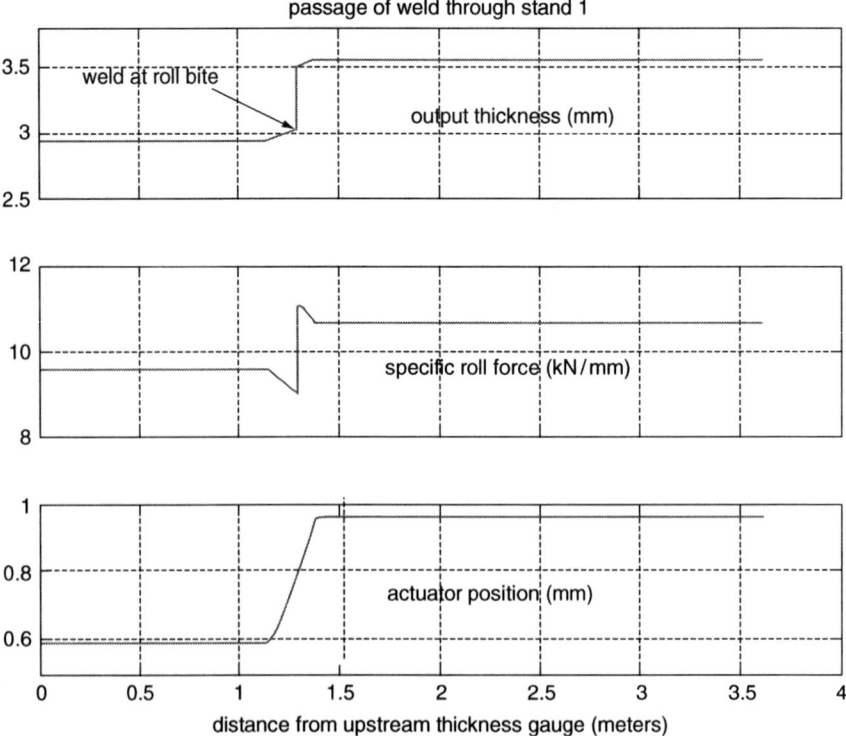

Fig. 5.25 Simulation results for path 2

a significant unmeasured disturbance that must be considered. To verify performance with this type of disturbance, simulation was done wherein a 10% increase in hardness was applied concurrently with the 10% increase in the thickness of the incoming material of the next strip. Table 5.20 depicts the results for this type of disturbance with the excursions in the listed variables as shown. Further increases in hardness can result in further significant increases in the excursions, particularly in the tension. If the excursions are excessive for the particular application, the use of the second method should be considered.

The measurement of strip speed, which ultimately affects the weld tracking function, is an uncertainty which could affect performance during the passage of the weld. Reliable high accuracy laser velocimeters typical of what is commercially available and used in operating mills are assumed to be used for this measurement, which has a typical uncertainty of 0.025% of measured strip speed. In addition the position measurement is recalibrated as the weld passes through each roll bite by a "weld in stand" signal. Based on this it was judged that the error in determining weld position was negligible and therefore could be excluded from the simulation.

Various other disturbances and uncertainties that are not peculiar to the weld passage and which could occur during tandem cold rolling are addressed during simulations of the stand-alone tandem cold rolling process.

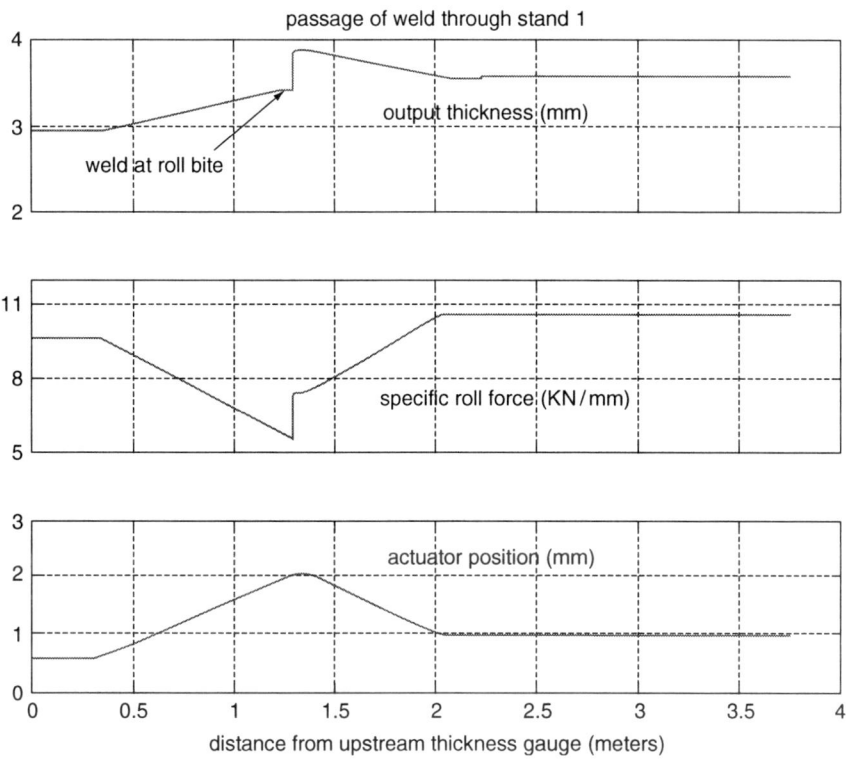

Fig. 5.26 Simulation results for path 5

Table 5.17 Simulation results for 5% thickness transition

	Stand 1	Stand 2	Stand 3	Stand 4	Stand 5
Out of tolerance length, m	0.14	0.13	0.18	0.28	0.27
Max change in specific roll force, kN/mm	0.43	0.31	0.31	0.31	0.19
Peak specific roll force	9.73	9.92	10.5	10.2	6.39
Max tension excursion, percent	−3.7	−3.4	−1.2	3.3	–

Table 5.18 Simulation results for 10% thickness transition

	Stand 1	Stand 2	Stand 3	Stand 4	Stand 5
Out of tolerance length, m	0.16	0.26	0.33	0.33	0.38
Max change in specific roll force, kN/mm	0.94	0.70	0.54	1.9	0.54
Peak specific roll force	9.88	10.3	10.7	11.3	6.59
Max tension excursion, percent	−7.7	−8.8	−1.1	−7.2	–

Table 5.19 Simulation results for 20% thickness transition

	Stand 1	Stand 2	Stand 3	Stand 4	Stand 5
Out of tolerance length, m	0.20	0.25	0.44	0.69	0.64
Max change in specific roll force, kN/mm	2.08	2.55	1.22	2.16	0.98
Peak specific roll force	11.2	12.2	11.2	11.8	6.86
Max tension excursion, percent	−11.0	−8.6	−9.4	−9.2	−

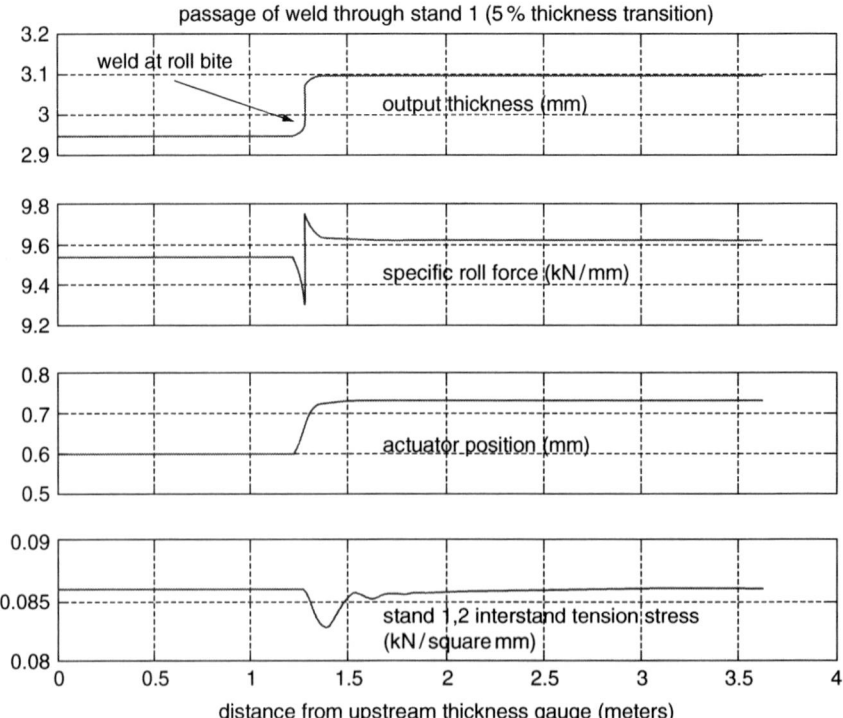

Fig. 5.27 Responses for 5% thickness transition

5.4.4.4 Mass Flow Considerations

Immediately after the transition has exited the roll bite of stand $i + 1$, the mass flow at the upstream stand was estimated by computing MF_{pctmax} as

$$MF_{pct\,max} = \max \left| \frac{MF_{tr,exit,i+1} - MF_{tr,exit,i}}{MF_{steady-state}} \right| 100\%, \qquad (5.58)$$

where $MF_{pct\,max}$ is the magnitude of the maximum percent deviation of the transient difference between the mass flows at the exits of stand $i + 1$ and stand i (*i.e.*, $MF_{tr,exit,i}$ $_{+1}$ and $MF_{tr,exit,i}$) with respect to the steady-state mass flow ($MF_{steady\text{-}state}$). $MF_{pct\,max}$ as determined by (5.58) is tabulated in Table 5.21.

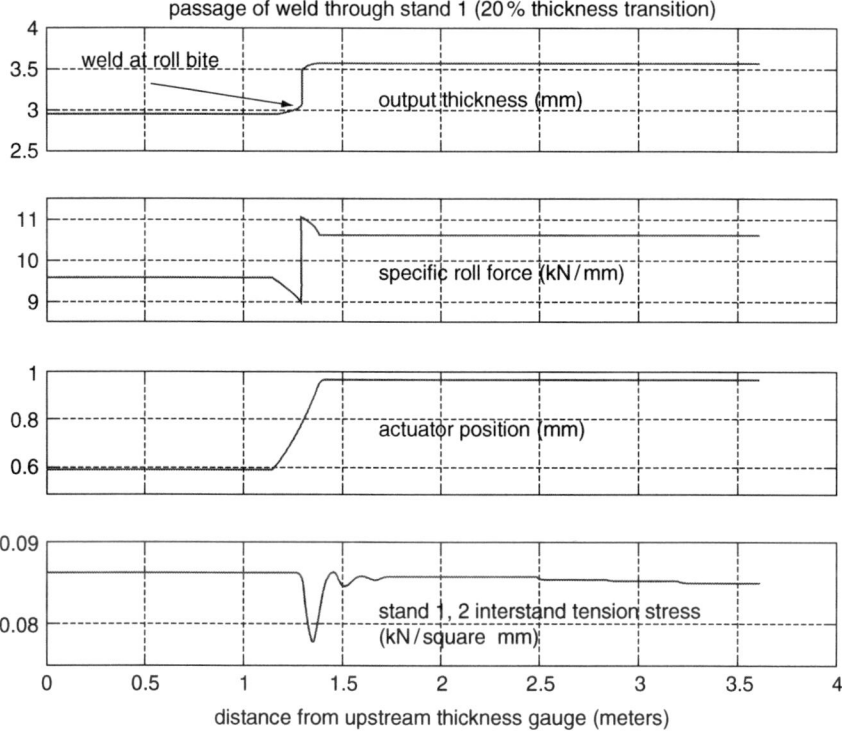

Fig. 5.28 Responses for 20% thickness transition

Table 5.20 Simulation results for a transition of 10% thickness increase and 10% hardness increase

	Stand 1	Stand 2	Stand 3	Stand 4	Stand 5
Out of tolerance length, m	0.25	0.58	0.86	0.76	0.76
Max change in specific roll force, kn/mm	1.76	1.73	1.57	3.22	0.98
Peak specific roll force	10.7	11.5	11.8	12.6	7.3
Max tension excursion, percent	32.0	34.0	−32.0	−32.0	−

Table 5.21 Simulation results for mass flow at stands i, $i+1$

Transition	Maximum magnitude of the percent deviation of the transient difference at the exits of			
	Stands 2,1	Stands 3,2	Stands 4,3	Stands 5,4
5%	0.16%	0.10	0.06	0.1
10	0.18	0.12	0.05	0.7
20	0.30	0.14	0.07	1.5

The maximums of the magnitudes of the percent deviations of the transient differences between the mass flows at other adjacent upstream stands (such as stands i and $i-1$) were determined similarly during passage of the transition

Table 5.22 Simulation results for mass flow at upstream stands (20% thickness transition)

Transition passing through	Maximum magnitude of the percent deviation of the transient difference at the exits of		
	Stands 2,1	Stands 3,2	Stands 4,3
Stand 3	0.14%	–	–
Stand 4	0.20	0.30	–
Stand 5	0.60	0.70	2.3

through the roll bite of stand $i+1$. The maximums of the magnitudes of these deviations for a thickness transition of 20% are as shown in Table 5.22. It also was determined that the maximums of the magnitudes of the deviations for thickness transitions of 5% and 10% were less than 0.1% and 0.4%, respectively. These results and those tabulated in Tables 5.21 and 5.22 show that the transient differences in the mass flows between adjacent upstream stands are not excessive. Additional simulations were performed which showed similar results for the transient differences in mass flows between non-adjacent upstream stands, and also confirmed that at steady-state there is zero difference between these mass flows. Similar results were obtained for the mass flows at downstream stands (such as stands $i+2$ and $i+3$) during passage of the transition through stand $i+1$.

5.4.5 *Comparison with Industrial Controllers*

An evaluation of the performance of the augmented state-dependent SDRE-based controller against actual industrial installations can be done only in general. This is because much of the data from operating industrial installations that could be useful for a more thorough comparison usually is unavailable. For example what often is available is the deviation from the desired centerline thickness at the mill output, and this is used as a basis for comparison. Other data which specifically describe the characteristics of disturbances, uncertainties, material properties such as material hardness, exit thicknesses at intermediate stands, and sometimes interstand tensions, are mostly unavailable. In addition, the actual mill characteristics will vary between different installations and can be different for the same installation depending on the operational conditions, *e.g.* the mill characteristics will change with mechanical wear and temperature which generally increase during mill usage. All of these considered together make it difficult (if even possible) to model disturbances and uncertainties in a simulated environment to very closely match conditions under which the output thickness data was collected.

In addition, it is also very difficult to model an actual installed controller and use it as a base for simulations against which the performance of a controller being developed can be evaluated. This is because much of the industrial controller structure is usually proprietary and very often the actual tuning settings also are proprietary or are never recorded during commissioning.

Table 5.23 Comparison of simulation results with industrial data

Controller	Magnitude of maximum percent deviation of mill output thickness
SDRE-based	0.2%
Industrial a [35]	0.5
Industrial b [36]	0.8

However, since the assumptions made for the simulations described previously are conservative, and the disturbances and uncertainties are combined in a manner to simulate the more severe conditions, it is considered that the results of these simulations can be taken to be conservative with respect to the actual mill data. On this basis Table 5.23 presents the mill exit thickness data for the augmented SDRE-based controller and two industrial controllers [35, 36]. The listings presented for the SDRE-based controller reflect the following:

- The maximum deviation in mill output thickness due to disturbances and uncertainties is about 0.08% based on the data listed in Tables 5.10–5.13.
- A deviation of about 0.05% due to cold mill roll eccentricity is assumed based on the results of simulations of a conceptual method of eccentricity compensation.
- A "safety factor" of 0.07% is assumed for changes in width and other effects.
- The total deviation in output thickness is taken as the sum of the above, *i.e.* about 0.2%.

As can be seen from the listings of Table 5.23 this compares well with data from two other well-performing industrial installations. Moreover, in addition to the improvement in reducing excursion in mill output thickness, the augmented SDRE-based controller has kept the excursions in the interstand tensions and output thicknesses at reduced levels which have contributed strongly to the stability of rolling. In the case of the continuous mill the magnitudes of the transient excursions in the mass flows during passage of the transition between the present strip and the next strip have been kept at very low levels which also contributed to the rolling stability.

5.5   Concluding Comments

The material presented in this chapter represents the work done to date for a method of advanced control of the tandem cold rolling process. The results of this effort have shown that the augmented SDRE technique offers a highly effective method of controlling thicknesses and tensions in a tandem cold metal rolling application. Good performance in the presence of realistic disturbances and uncertainties has been demonstrated for various regimes of mill operation for both the stand-alone case and the continuous case. Moreover, the augmented SDRE control structure which includes the trims provides a user-friendly environment for the control designer and the commissioning engineer. The work performed represents what was accomplished with the resources available for support and provides a strong indication of the success of any future efforts. While the results have demonstrated a highly successful and improved performance, recommendations for the support of future activity to provide even further improvements include the investigation of the possibility of the deletion of the strip speed sensors without degrading performance, the expansion of the simulation to confirm that performance can be retained for the full scope of products that are expected to be processed over a broad range of applications, control during threading, and eventual application to an actual mill. The work described in this section gives evidence that such an application has a high likelihood of success.

Appendix

The sections of this appendix provide material that supplements what is presented earlier in this chapter. Provided are some applicable definitions and theorems, detailed methods for the computation of gradients, derivations of various relationships relating to functions of the state variables, and a derivation of the necessary conditions for the optimality of the SDRE controller.

Definitions and Theorems

The definitions and theorems which follow are related to the material presented in Section 5.3.1. The definitions are based on [37] and [15]. Theorems are based on [15]. A more detailed theoretical treatment is found in [12] and the references cited therein.

Definition 1. *A system is considered an **autonomous system** if the function f does not depend explicitly on t, i.e.*

$$\dot{x} = f(x). \tag{5.59}$$

Definitions 2 through 8 are based on an autonomous system, where $f : D \rightarrow R^n$ is a locally Lipshitz map from D into R^n.
Definition 2. *The point $\tilde{x}$ is an **equilibrium point** of (5.59) if*

$$f(\tilde{x}) = 0. \tag{5.60}$$

Definition 3. *Taking $\tilde{x} = 0$ for convenience and without loss of generality, the equilibrium point of (5.60) is **stable** if for each $\varepsilon > 0$ there is a δ such that*

$$\|x(0)\| < \delta \Rightarrow \|x(t)\| < \varepsilon, \ \forall t \geq 0. \tag{5.61}$$

Definition 4. *The equilibrium point of (5.60) is* **unstable** *if it is not stable.*

Definition 5. *The equilibrium point* $\tilde{x} = 0$ *is* **asymptotically stable** *if it is stable and a δ can be chosen such that*

$$\|x(0)\| < \delta \Rightarrow \lim_{t \to \infty} x(t) = 0. \tag{5.62}$$

Definition 6. *Let $\phi(x; t)$ be the solution of (5.59) that starts at time $t = 0$ and at an initial state x_0, with $\tilde{x} = 0$. Then the* **region of attraction** *is the set of all points x such that*

$$\lim_{t \to \infty} \phi(x; t) = 0. \tag{5.63}$$

Definition 7. *The equilibrium point $\tilde{x} = 0$ is* **locally asymptotically stable** *if it is asymptotically stable and its region of attraction is some neighborhood of the origin.*

Definition 8. *The equilibrium point $\tilde{x} = 0$ is* **globally asymptotically stable** *if*

$$\lim_{t \to \infty} \phi(x; t) = 0, \tag{5.64}$$

no matter how large $\|x\|$ is.

Definition 9. *$\{C(x), A(x)\}$ is a* **pointwise observable** *parameterization of the nonlinear system in a region Ω if the pair $\{C(x), A(x)\}$ is pointwise observable (in the linear sense) for all $x \in \Omega$.*

Definition 10. *$\{C(x), A(x)\}$ is a* **pointwise detectable** *parameterization of the nonlinear system in a region Ω if the pair $\{C(x), A(x)\}$ is pointwise detectable[5] (in the linear sense) for all $x \in \Omega$.*

Definition 11. *$\{A(x), B(x)\}$ is a* **pointwise controllable** *parameterization of the nonlinear system in a region Ω if the pair $\{A(x), B(x)\}$ is pointwise controllable (in the linear sense) for all $x \in \Omega$.*

Definition 12. *$\{A(x), B(x)\}$ is a* **pointwise stabilizable** *parameterization of the nonlinear system in a region Ω if the pair $\{A(x), B(x)\}$ is pointwise stabilizable[6] (in the linear sense) for all $x \in \Omega$.*

Theorem 1. *In addition to $a(x), b(x), R(x), Q(x), Q(x) \in C^k, k \geq 1$, assume that $A(x)$ is smooth (i.e. $A(x) \in C^k$) and that $A(x)$ is both a stabilizable and detectable coefficient parameterization of the nonlinear system. Then the state-dependent*

---

[5]In a linear sense, the pair $\{C,A\}$ is detectable if and only if every unstable mode is observable [38].

[6]In a linear sense, the pair $\{A,B\}$ is stabilizable if and only if every unstable mode is controllable [38].

Riccati equation method produces a closed-loop solution which is locally asymptotically stable.

Proof: The proof is provided in [15].

Theorem 2. *Assume that the functions A(x), b(x), K(x), Q(x), and R(x), and their gradients*[7] *$\nabla_x A(x)$, $\nabla_x b(x)$, $\nabla_x K(x)$, and $\nabla_x A(x)$ are bounded along trajectories. Then, under stability, as the state x is driven to zero, the necessary condition for optimality is asymptotically satisfied at a quadratic rate.*

Proof: The proof is provided in [15].

Computation of Gradients

With $x \in R^n$, $Q'(x) = Q(x) \in R^{n \times n}$, and $Q(x) \in C^1$, and using matrix differentiation formulae as given in [39],

$$\nabla_x(x'Q(x)x) = 2Q(x)x + x'\nabla_x Q(x)x, \qquad (5.65)$$

where

$$\nabla_x(x'Q(x)x) = \begin{bmatrix} x'\nabla_{x1}Q(x)x \\ x'\nabla_{x2}Q(x)x \\ \vdots \\ x'\nabla_{xn}Q(x)x \end{bmatrix}, \qquad (5.66)$$

and

$$\nabla_{xi}Q(x) = \begin{bmatrix} \frac{\partial q_{11}(x)}{\partial x_i} & \frac{\partial q_{12}(x)}{\partial x_i} & \cdots & \frac{\partial q_{1n}(x)}{\partial x_i} \\ \vdots & & & \vdots \\ \frac{\partial q_{n1}(x)}{\partial x_i} & \cdots & \cdots & \cdots & \frac{\partial q_{nn}(x)}{\partial x_i} \end{bmatrix}, \quad i = 1, 2, \ldots n. \qquad (5.67)$$

Equation 5.65 can be verified by first computing $x'Q(x)x$ on an element-by-element basis, and then computing $\nabla_x(x'Q(x)x)$.

Example 1. *Computation of $\nabla_x(x'Q(x)x)$, for $x \in R^2$, $Q'(x) = Q(x) \in R^{2 \times 2}$, $Q(x) \in C^1$:*

Using (5.65) and (5.66), and not showing function arguments,

---

[7]The notation for gradient is as given in Section "Computation of Gradients" in the Appendix.

$$\nabla_x(x'Qx) = 2 \begin{bmatrix} q_{11} & q_{12} \\ q_{21} & q_{22} \end{bmatrix} \begin{bmatrix} x_1 \\ x_2 \end{bmatrix} + \begin{bmatrix} x'\nabla_{x1}Qx \\ x'\nabla_{x2}Qx \end{bmatrix}, \tag{5.68}$$

and then using (5.67),

$$x'\nabla_{x1}Qx = [x_1 \ \ x_2] \begin{bmatrix} \dfrac{\partial q_{11}(x)}{\partial x_1} & \dfrac{\partial q_{12}(x)}{\partial x_1} \\[2ex] \dfrac{\partial q_{21}(x)}{\partial x_1} & \dfrac{\partial q_{22}(x)}{\partial x_1} \end{bmatrix} \begin{bmatrix} x_1 \\ x_2 \end{bmatrix}, \tag{5.69}$$

$$x'\nabla_{x2}Qx = [x_1 \ \ x_2] \begin{bmatrix} \dfrac{\partial q_{11}(x)}{\partial x_2} & \dfrac{\partial q_{12}(x)}{\partial x_2} \\[2ex] \dfrac{\partial q_{21}(x)}{\partial x_2} & \dfrac{\partial q_{22}(x)}{\partial x_2} \end{bmatrix} \begin{bmatrix} x_1 \\ x_2 \end{bmatrix}. \tag{5.70}$$

Performing the multiplications and substituting into (5.68), and noting that $q_{12} = q_{21}$,

$$\nabla_x(x'Qx) = \begin{bmatrix} 2x_1q_{11} + 2x_2q_{12} + x_1^2\dfrac{\partial q_{11}}{\partial x_1} + 2x_1x_2\dfrac{\partial q_{12}}{\partial x_1} + x_2^2\dfrac{\partial q_{22}}{\partial x_1} \\[2ex] 2x_1q_{12} + 2x_2q_{22} + x_1^2\dfrac{\partial q_{11}}{\partial x_2} + 2x_1x_2\dfrac{\partial q_{12}}{\partial x_2} + x_2^2\dfrac{\partial q_{22}}{\partial x_2} \end{bmatrix}, \tag{5.71}$$

Computing $x'Qx$ *on an element-by-element basis and then computing* $\nabla_x(x'Qx)$ *verifies the result obtained in (5.71).*

Example 2. *Computation of* $\nabla_x(\lambda'A(x)x)$, *where* $x \in R^2$, $A(x) \in R^{2 \times 2}$, $A(x) \in C^1$, $\lambda \in R^2$, *for all* x:

Using matrix differentiation formulae, and not showing function arguments,

$$\nabla_x(\lambda'Ax) = (x'\nabla_xA' + A')\lambda, \tag{5.72}$$

where

$$x'\nabla_xA' = \begin{bmatrix} x'\nabla_{x1}A' \\ x'\nabla_{x2}A' \end{bmatrix}, \tag{5.73}$$

and

$$x'\nabla_{x1}A = [x_1 \ \ x_2] \begin{bmatrix} \dfrac{\partial a_{11}}{\partial x_1} & \dfrac{\partial a_{12}}{\partial x_1} \\[2ex] \dfrac{\partial a_{21}}{\partial x_1} & \dfrac{\partial a_{22}}{\partial x_1} \end{bmatrix}, \tag{5.74}$$

$$x'\nabla_{x2}A = \begin{bmatrix} x_1 & x_2 \end{bmatrix} \begin{bmatrix} \dfrac{\partial a_{11}}{\partial x_2} & \dfrac{\partial a_{12}}{\partial x_2} \\[2ex] \dfrac{\partial a_{21}}{\partial x_2} & \dfrac{\partial a_{22}}{\partial x_2} \end{bmatrix}. \tag{5.75}$$

It is straightforward to use (5.74) and (5.75) and substitute into (5.72) and (5.73) to obtain the result.

Derivation of Relationships as Functions of the State Variables

The material which follows relates closely to the determination of the state-dependent elements of the $A(x)$ matrix as noted in Section 5.3.3.

Relationships for Output Thickness and Specific Roll Force

During each scan of the controller, ξ and α are computed at a number of equally spaced points in a predetermined neighborhood of h_{out0} as

$$\xi = \frac{\mu\sqrt{R'(h_{in} - h_{out})}}{\bar{h}}, \tag{5.76}$$

$$\alpha = \sqrt{\frac{h_{out}}{h_{in}}}\exp(\xi) - 1, \tag{5.77}$$

where h_{in} for stand 1 is the input thickness to the mill, and for stands 2,3,4,5 is the output thickness of the previous stand delayed by the appropriate interstand time delay,

$$h_{in,i+1}(t) = h_{out,i}(t - t_{d,i,i+1}), \quad i = 1, 2, 3, 4, \tag{5.78}$$

where μ and R' are the friction coefficient and the deformed work roll radius, and $t_{d,i,i+1}$ is the time delay between stands i and $i + 1$.

During the same controller scan, using the relationship (2.12) for the specific roll force and noting that $F = PW$, the total roll force is computed (at each point) as

$$F = (\bar{k} - \bar{\sigma})\sqrt{R'\delta}(1 + 0.4\alpha)W, \tag{5.79}$$

where W is the strip width and other variables are as denoted in Chapter 2. In the neighborhood of h_{out0}, F then is approximated by a linear fit which is reasonable because the neighborhood is not large,

$$F = c_1 h_{out} + c_2, \qquad (5.80)$$

where c_1 and c_2 are constants. Using (2.26) and (5.80), h_{out} is then

$$h_{out} = \frac{M(S + S_0)}{(M - c_1)}, \qquad (5.81)$$

and the specific roll force is

$$P = \frac{M(h_{out} - (S + S_0))}{W}, \qquad (5.82)$$

and thus for stands 2,3,4,5, h_{out} and P depend on the state variables.

Relationships for Entry and Exit Strip Speeds

Using (2.19) the strip speed at the exit of the roll bite can be written as

$$V_{out} = V(f + 1), \qquad (5.83)$$

where the forward slip f as given in (2.23) ultimately depends on the state variables. By conservation of mass flow across the roll bite,

$$V_{in.i+1} = \frac{V_{out,i+1} h_{out.i+1}}{h_{in,i+1}} = \frac{V_{i+1}(f_{i+1} + 1) h_{out,i+1}}{h_{out,i}(t - t_{d.i,i+1})}, \qquad (5.84)$$

and thus $(V_{in,i+1} - V_{out,i})$ also depends on the state variables.

Derivation of the Necessary Conditions for Optimality

What is presented in this section supports the material discussed in Section 5.3.1.

From the cost function (5.17) and the nonlinear constraint (5.15), the Hamiltonian function is formed as:

$$H(x, u, \lambda) = \tfrac{1}{2}(x'Q(x)x + u'R(x)u) + \lambda'(A(x)x + Bu), \qquad (5.85)$$

where $\lambda \in R^n$ is a Lagrange multiplier. The necessary conditions for optimality of a nonlinear controller are:

$$\nabla_\lambda H = \dot{x}, \qquad (5.86)$$

$$\nabla_x H = -\dot{\lambda}, \qquad (5.87)$$

$$\nabla_u H = 0. \tag{5.88}$$

Using (5.85) and the control law (5.19),

$$u = -R^{-1}(x)B'K(x)x, \tag{5.89}$$

$$\nabla_u H = R(x)u + B'\lambda, \tag{5.90}$$

$$\nabla_u H = R(x)(-R^{-1}(x)B'K(x)x) + B'\lambda, \tag{5.91}$$

$$\nabla_u H = B'(\lambda - K(x)x). \tag{5.92}$$

$\nabla_u H$ will be zero if λ is chosen so that

$$\lambda = K(x)x. \tag{5.93}$$

Differentiating with respect to time results in

$$\dot{\lambda} = \dot{K}(x)x + K(x)\dot{x}. \tag{5.94}$$

Using (5.85) and (5.87),

$$\dot{\lambda} = -Q(x)x - \tfrac{1}{2}(x'\nabla_x Q(x)x + u'\nabla_x R(x)u) - (x'\nabla_x A'(x) + A'(x))\lambda. \tag{5.95}$$

Equating (5.94) and (5.95), and using the nonlinear constraint (5.15), (5.86), (5.89), and (5.93),

$$\dot{K}(x)x + K(x)(A(x)x - BR^{-1}(x)B'K(x)x) =$$
$$- Q(x)x - \tfrac{1}{2}(x'\nabla_x Q(x)x + u'\nabla_x R(x)u) - (x'\nabla_x A'(x) + A'(x))K(x)x. \tag{5.96}$$

Rearranging and grouping terms,

$$\dot{K}(x)x + \tfrac{1}{2}(x'\nabla_x Q(x)x + u'\nabla_x R(x)u) + x'\nabla_x A'(x)K(x)x$$
$$+ (A'(x)K(x) + K(x)A(x) - K(x)BR^{-1}(x)B'K(x) + Q(x))x = 0. \tag{5.97}$$

From the state-dependent algebraic Riccati equation, the expression $(A'(x) K(x)+K(x) A(x) - K(x) B R^{-1}(x) B' K(x) + Q(x))$ is equal to zero, and substituting for u (5.89) gives the necessary condition for the closed-loop solution to be near-optimal

$$\dot{K}(x)x + \tfrac{1}{2}(x'\nabla_x Q(x)x + x'K(x)BR^{-1}(x)\nabla_x R(x)R^{-1}(x)B'K(x)x)$$
$$+ x'\nabla_x A'(x)K(x)x = 0. \tag{5.98}$$

References

1. Pearson JD. Approximation methods in optimal control. J Electron Control. 1962;13:453–69.
2. Wernli A, Cook G. Suboptimal control for the nonlinear quadratic regulator problem. Automatica. 1975;11:75–84.
3. Mracek CP, Cloutier JR. Control designs for the nonlinear benchmark problem via the state-dependent Riccati equation method. Int J Robust Nonlinear Control. 1998;8:401–33.
4. Friedland B. Advanced control system design. Englewood Cliffs: Prentice-Hall; 1996.
5. Parrish DK, Ridgely DB. Control of an artificial human pancreas using the SDRE method. In: Proceedings of American Control Conference, Albuquerque; 1997.
6. Cloutier JR, Stansbery DT. Nonlinear, hybrid bank-to-turn/skid-to-turn autopilot design. Proceedings of the AIAA Guidance, Navigation, and Control Conference; 2001; Montreal.
7. Cloutier JR, Zipfel PH. Hypersonic guidance via the state-dependent-Riccati equation control method. Proceedings of IEEE Conference of Control Applications; 1991; Hawaii.
8. Stansbery DT, Cloutier JR. Position and attitude control of a spacecraft using the state-dependent-Riccati equation technique. Proceedings of American Control Conference; 2000; Chicago.
9. Beeler SC, Kepler GM. Reduced order modeling and control of thin film growth in an HPCVD reactor. Control Research Computation Report CRSC-TR00-33. Raleigh: North Carolina State University; 2000.
10. Athans M, Falb PL. Optimal control an introduction to the theory and its applications. New York: McGraw-Hill; 1966.
11. Lewis FL. Optimal control. New York: Wiley; 1986.
12. Cimen T. State-dependent Riccati equation (SDRE) control: a survey. In: Chung MJ, Misra P, editors. Plenary papers, milestone reports, and select survey papers. Seoul: 17th IFAC World Congress; 2008.
13. Erdem EB. Analysis and real-time implementation of state-dependent Riccati equation controlled systems, PhD thesis. Champaign: University of Illinois at Urbana-Champaign; 2001.
14. Hammet KD. Control of nonlinear systems via state feedback state-dependent Riccati equation techniques. PhD thesis, Air Force Institute of Technology, Ohio, 1997.
15. Coutier JR, D'Souza N, Mracek CP. Nonlinear regulation and nonlinear H^∞ control via the state-dependent Riccati equation technique: part 1, theory. Proceedings of International Conference on Nonlinear Problems in Aviation and Aerospace. Dayton Beach: Embry Riddle University; 1996; P. 117–31.
16. Vidyasagar M. Nonlinear systems analysis. Englewood Cliffs: Prentice-Hall; 1978.
17. Tsiotras P, Corless M, Rotea M. Counterexample to a recent result on the stability of nonlinear systems. IMA J Math Control Info. 1996;13(2):129–30.
18. Pittner J. Pointwise linear quadratic control of a tandem cold rolling mill, PhD thesis. Pittsburgh: University of Pittsburgh; 2006.
19. Pittner J, Simaan MA. Optimal control of tandem cold rolling using a pointwise linear quadratic technique with trims. J Dyn Syst-T ASME. 2008;130(2):021006-1–021006-11.
20. Kinney CS, Laub AJ. The matrix sign function. IEEE Trans Automat Control. 1995;40(8): 1330–48.
21. Kugi A, Haas W, Schlacher K, et al. Active compensation in rolling mills. IEEE Trans Ind Appl. 2000;36(2):625–32.
22. Hayes MH. Statistical digital signal processing and modeling. New York: Wiley; 1996.
23. Widrow B, Streans S. Adaptive signal processing. Englewood Cliffs: Prentice-Hall; 1986.
24. Neumerkel D, Shorten R, Hambrecht A. Robust learning algorithms for nonlinear filtering. IEEE International Conference on Acoustics, Speech, Signal Processing. vol 6; Atlanta 1996; P. 3565–58.
25. Rawlings JB. Tutorial overview of model predictive control. IEEE Control Syst Mag. 2000;20 (3):38–52.

26. Allgower F, Badgwell TA, Qin SJ, et al. Nonlinear predictive control and moving horizon estimation-an introductory overview. In: Frank PM, editor. Advances in control: highlights of ECC99. London: Springer; 1999.
27. Edwards E, Spurgeon SK. Sliding mode control theory and applications. London: Taylor & Francis; 1998.
28. Slotine JJE, Li W. Applied nonlinear control. Englewood Cliffs: Prentice-Hall; 1991.
29. Pittner J, Simaan MA. Optimum feedback controller design for tandem cold metal rolling. Proceedings of IFAC 17th World Congress; 2008; Seoul, P. 988–93.
30. Pittner J, Simaan MA. Optimal control of continuous tandem cold metal rolling. In: Proceedings of American Control Conference; 2008; Seattle, P. 2018–23.
31. Pittner J, Simaan MA. Control of a continuous tandem cold rolling process. Control Eng Pract. 2008;16(11):1379–90.
32. Pittner J, Samaras NS, Simaan MA. A new strategy for optimal control of continuous tandem cold metal rolling. IEEE Trans Ind Appl. 2010;46(2):703–11.
33. Roberts WL. Cold rolling of steel. New York: Marcel Dekker; 1978.
34. Ginzburg V, Ballas R. Roll eccentricity. United Engineering. Pittsburgh: International Rolling Mill Consultants. Pittsburgh; 1998.
35. Tezuka T, Yamashita T, Sato T, et al. Application of a new automatic gauge control system for the tandem cold mill. IEEE Trans Ind Appl. 2002;38(2):553–8.
36. Sekiguchi K, Seki Y, Okitani N, et al. The advanced set-up and control system for Dofasco's tandem cold mill. IEEE Trans Ind Appl. 1996;32(3):608–616.
37. Khalil HK. Nonlinear systems. New York: Macmillan; 1992.
38. Zhou K, Doyle JC. Essentials of robust control. Upper Saddle River: Prentice-Hall; 1998.
39. Vetter W. Derivative operations on matrices. IEEE Trans Automat Control. 1970;15(2): 241–4.

Chapter 6
Main Drives and Motors

6.1 Introduction

The objective of the material provided in this chapter is to supplement what is provided in previous chapters to present a more fully rounded-out picture of the control of tandem cold metal rolling. The material of this chapter provides a very basic introduction to variable speed main drives and motors as applied to tandem cold rolling mills, so that the reader is provided with a background that will enable him or her to better understand the more complex issues associated with tandem cold mill main drives and motors as presented in various texts, archival literature, and manufacturer's publications. The main focus will be on voltage source converters in the medium voltage range as drives for synchronous motors and cage rotor induction motors as these are the most typical for modern applications. Some fundamentals related to these types of motors will be presented, with consideration as to their usage in variable speed applications such as the tandem cold rolling mill. Similarly, some basic concepts of the voltage source converter will be addressed along with the concept of closed-loop field oriented (*i.e.*, vector) control. Beneficial characteristics available in modern drives and motors also will be presented.

6.2 The Multi-phase Cage Rotor Induction Motor

The multi-phase cage rotor induction motor has seen wide usage in an almost endless variety of industrial applications where a simple, rugged, economical and nearly constant speed actuator is desired. The induction motor has a rotor and a stator which are separated by an air gap. The rotor is mounted on bearings, while the stator is fixed. Alternating currents exist in both the stator and the rotor. The currents in the stator are furnished by a source to which the stator is directly connected, while the currents in the rotor are the result of induction, hence the name induction motor. Most of the multi-phase motors in use are three phase, although in certain applications other multi-phase configurations are more desirable. It should be noted that the induction motor also is capable for operation as a

J. Pittner and M.A. Simaan, *Tandem Cold Metal Rolling Mill Control*,
Advances in Industrial Control, DOI 10.1007/978-0-85729-067-0_6,
© Springer-Verlag London Limited 2011

generator, and during certain operational conditions in a tandem cold rolling application can provide regenerative power to an external entity which could be other drive motors in the mill or coilers, or ultimately the plant power grid. This is addressed later in Section 6.2.3.

The stator frame of the motor is constructed of laminations of sheet steel that are of a special grade for use in motors. Slots are cut on the inside surface of the stator frame into which is put a three phase winding. The cage-type rotor is constructed similarly with slots cut into its outer surface into which are inserted conductive bars of aluminum or copper, both ends of which are short circuited by end rings. In a wound rotor machine the rotor winding is inserted into the slots cut into the rotor. In the cage rotor construction, there is no external connection of the rotor circuit to the outside world, as there is in the wound rotor construction wherein the rotor circuit is accessible via mechanical brushes and slip rings. The construction for the cage rotor type motor is depicted functionally in Figure 6.1, where the three phase winding of the stator is shown as three coils, denoted as $a\text{-}a'$, $b\text{-}b'$, and $c\text{-}c'$ each of which is excited by one phase of a balanced three phase source. The coils are depicted as three concentrated coils for simplicity. In actuality the three phase winding is distributed so as to produce a nearly sinusoidal spatial distribution of magnemotive force (mmf). More specifics regarding the fundamentals of induction motors can be found in reference texts on electric machinery, *e.g.*, [1, 2].

6.2.1   The Rotating Magnetic Field

To understand the operation of the multi-phase induction motor it is necessary to first understand how a rotating magnetic field is produced by the stator windings when connected to a balanced three phase source. In the three phase machine as depicted in Figure 6.1, the windings of the stator are displaced from each other

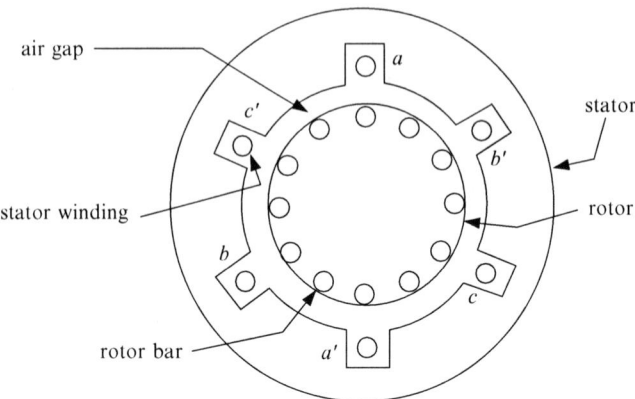

Fig. 6.1 Schematic of cage rotor and stator

around the air gap by 120° in a two-pole machine, which for a two-pole machine is also the electrical angular separation. If the winding pattern of the stator of Figure 6.1 were stretched out linearly in space the pattern as depicted in Figure 6.2 would result. The electrical angular separation can be understood by noting that the electrical angle θ_e is based on one cycle of flux distribution as shown in Figure 6.2, where one cycle is 360 electrical degrees. If the machine has P poles, then there are $P/2$ wavelengths per revolution so that the spatial (or mechanical) angle θ_m as measured in the physical degrees of rotation around the stator is

$$\theta_m = \frac{2}{P}\theta_e. \tag{6.1}$$

As an example in a four-pole machine the electrical angle is 120° which corresponds to 60° of spatial rotation or a spatial angle of 60°. In a two-pole machine the electrical and spatial angles are equal to each other.

Figure 6.2 depicts the mmf wave in a two-pole machine, which is produced by a fixed (*i.e.*, a time-invariant) current in coil a-a', around the air gap of the machine shown in Figure 6.1. The fixed current is into the page at a' and out of the page at a as indicated by the x and the dot shown respectively on the legs of the coil a. The peak of the mmf wave $F_{a(peak)}$ is proportional to the magnitude of the current in coil a. Coils b and c are located on the machine stator such that similar mmf waves are produced that are displaced at angles of 120° and 240° around the air gap from the mmf wave produced by the current in coil a. The angle θ is measured counterclockwise around the air gap that is depicted in Figure 6.1. The three mmf waves combine to give a total mmf wave, which can be described as

$$F(\theta) = F_{a(peak)}cos(\theta) + F_{b(peak)}cos(\theta - 120°) + F_{c(peak)}cos(\theta - 240°). \tag{6.2}$$

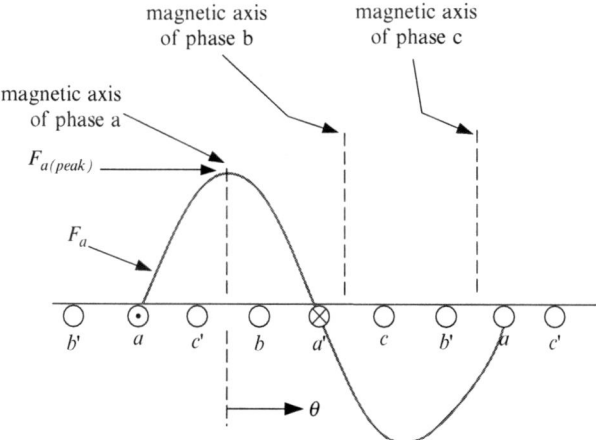

Fig. 6.2 Magnetomotive force wave of phase a

When the individual windings are excited by a balanced three phase source, the instantaneous peak mmf in a winding is proportional to the instantaneous current in the winding, so that the peaks of the mmfs in each winding are

$$F_{a(peak)} = F_{a(max)} cos(\omega t),$$ (6.3)

$$F_{b(peak)} = F_{b(max)} cos(\omega t - 120°),$$ (6.4)

$$F_{c(peak)} = F_{c(max)} cos(\omega t - 240°),$$ (6.5)

where ω is the frequency (radians/second) of the sinusoidal alternating current in a winding. Since the three phase source is balanced the peak current is the same in each of the phases, and since the number of turns in each of the windings is the same, the peaks of the mmfs are the same, *i.e.*,

$$F_{a(max)} = F_{b(max)} = F_{c(max)} = NI_{max},$$ (6.6)

where N is the number of winding turns and I_{max} is the peak of the alternating current, so that (6.2) becomes

$$F(\theta, t) = NI_{max} \; cos(\omega t)cos(\theta)$$
$$+ NI_{max} \; cos(\omega t - 120°) \; cos(\theta - 120°)$$
$$+ NI_{max} \; cos(\omega t - 240°) \; cos(\theta - 240°).$$ (6.7)

Using the trigonometric identity for the product of two cosines, (6.7) becomes

$$F(\theta, t) = \frac{3}{2} NI_{max} \; cos(\theta - \omega t),$$ (6.8)

which represents the rotation of the mmf wave around the air gap at the constant angular velocity ω, so that at any instant of time t_1 the positive peak of the wave is at an electrical angle of $\theta_1 = \omega t_1$, and later at time t_2 the wave has moved further around the air gap so that the positive peak is at $\theta_2 = \omega t_2$, and the wave has traveled over an electrical angle of $\theta_2 - \theta_1 = \omega (t_2 - t_1)$. The rotating wave also may be represented as an mmf vector rotating at angular velocity ω (electrical radians per second) around the air gap.

The above analysis is for a two-pole machine. In the case of higher order poles, the two-pole configuration is extended by locating the two-pole configuration serially around the stator of the machine, with the spatial angle reduced from the two-pole case according to the number of poles of the machine, as noted previously. For any machine the rotational velocity ω_m of the mmf wave in radians/second with respect to the actual stator is

$$\omega_m = \frac{2}{P} \omega_e, \text{or } n = \frac{120f}{P}, \text{in rev/min},$$ (6.9)

where P is number of poles of the machine, and f is the frequency of the three phase source in Hertz. This rotational velocity is denoted as the synchronous speed of the machine. The per-unit difference in the speed of the rotor with respect to the synchronous speed is the per-unit slip which is expressed as

$$s = \frac{n_1 - n}{n_1}, \tag{6.10}$$

where n_1 is the synchronous speed and n is the actual rotor speed. The frequency of the currents in the rotor are then expressed in terms of the slip as $f_r = sf$, where f_r is the slip frequency, with f being the frequency of the stator currents. At starting the rotor is stationary and the slip is 1.0, with the frequency of the rotor currents equal to the frequency of the stator currents.

6.2.2 Machine Torque

An understanding of how the rotor mmf and air gap flux-density waves of the machine interact to produce torque also is essential to understanding the operation of the cage rotor induction motor. As a starting point, a cage rotor machine that is powered by a balanced three phase source will be considered, wherein the rotor is being driven by an external motor at exactly the synchronous speed of the machine. In this case there is a rotating flux-density wave that is produced by the rotating mmf wave around the air gap of the machine and it is assumed that magnetic saturation effects are negligible so that the rotating mmf wave produced by the stator can be expressed in terms of a rotating flux-density wave that rotates in phase (*i.e.*, at the same angle) as the mmf wave. Since the rotor is being turned at synchronous speed, there are no voltages induced in the rotor bars since the rate of change of flux linking the rotor bars is zero.

If next the external motor driving the rotor is instantly decoupled from the rotor shaft, the rotor will begin to slow down due to the decelerating torque caused by the machine losses. As the rotor begins to decelerate, the rate of change of flux in the rotor bars is no longer zero and the changing flux induces voltages in the bars. These voltages produce currents in the bars which in turn generate a rotor mmf wave that rotates at the same speed as the flux-density wave and interacts with it to produce a torque that eventually balances the torque caused by the machine losses. Since the torque required to balance the losses is quite low, the rotor is turning almost at synchronous speed and thus the rate of change of flux in the rotor bars also is quite low, and therefore the frequency of the induced voltages and the resulting rotor bar currents, is very low. Referring to Figure 6.3, since the frequency of these induced currents I_r is very low, the rotor reactance X_r also is very low so that the rotor impedance is essentially resistive (*i.e.*, nearly equal to R_r) and thus the rotor currents are essentially in phase with the voltages E_r induced in the rotor. The resulting mmf wave produced by the rotor currents thus lags behind the flux-density

Fig. 6.3 Rotor circuit

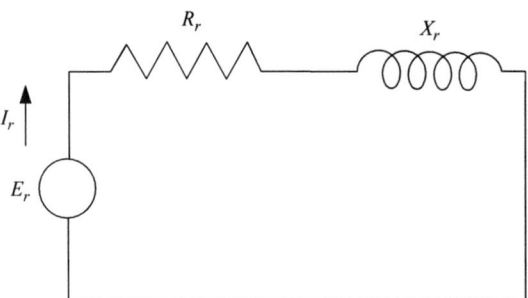

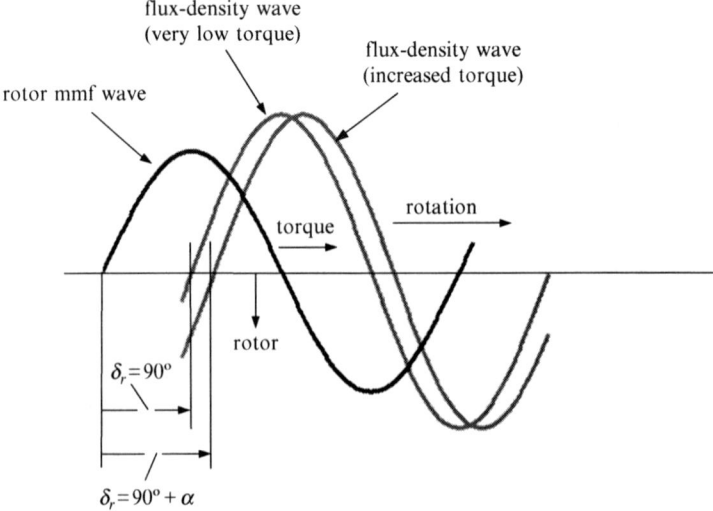

Fig. 6.4 Change in position of flux-density wave as load torque is increased

wave by slightly more than 90 electrical degrees and is no longer produced only by the stator mmf, but is the result of both the stator mmf and the rotor mmf.

If now some external load torque is applied to the rotor shaft, the rotor will begin to slow further to produce torque that balances the load torque, the frequency of the rotor currents will increase and thus the rotor reactance will increase which will cause an increase in the angle between the rotor induced voltage and the rotor currents, with a resulting increase of α in the angle δ_r between the flux-density wave and the rotor mmf wave as depicted in Figure 6.4, where for simplicity any changes in the peaks of the waves due to application of the increased load torque are not shown.

If the load torque is increased yet further, the rotor speed will decrease further and eventually the rotor reactance will increase to the point where the angle between the rotor mmf wave and the flux-density wave will be such that the torque decreases rather than increases. The torque at this point is denoted as the breakdown torque, which is the maximum torque that can be produced by the motor at its nominal (or base) frequency and rated terminal voltage. The breakdown torque is greater than the rated torque of the machine and limits the short time overload capacity of the motor. A typical torque-speed curve for a cage rotor induction motor at its nominal frequency and rated terminal voltage is depicted in Figure 6.5, with the rated torque, breakdown torque, stall torque and pull-up torque shown. The stall torque is the torque developed with the rotor locked, the pull-up torque is the minimum torque available to accelerate the rotor and any coupled load.

An expression for the rotor torque T can be derived [1] as

$$T = K_T \left(\frac{P}{2}\right)^2 B_{sr} F_r \sin(\delta_r), \tag{6.11}$$

where P is the number of poles of the machine, B_{sr} is the peak value of the flux-density wave, F_r is the peak of the mmf wave resulting from the rotor currents, δ_r is the electrical angle between the two waves, and K_T is a constant depending on rotor geometry. Equation 6.11 can be approximated by (6.12) as B_{sr} is approximately constant and F_r is proportional to the root-mean-square of the rotor current,

$$T = K I_r \sin(\delta_r), \tag{6.12}$$

where K is a constant.

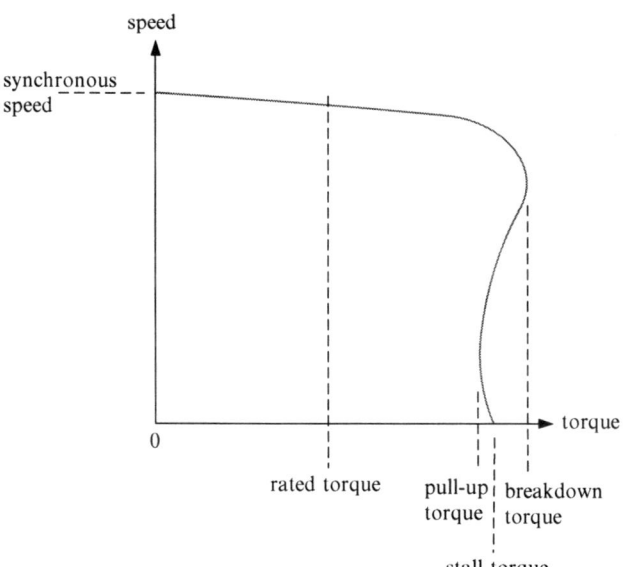

Fig. 6.5 Typical induction motor flux-density characteristic, motoring

6.2.3 Regenerative Operation

In a tandem cold rolling operation a cage rotor induction motor can regenerate power
to other motors that control the mill or to the plant power grid via the drive system
associated with the motor. The regeneration of power during steady speed mill
operation can occur at an unwind motor of a stand-alone mill to maintain tension
at the mill entry. During a decrease in the mill speed, certain of the mill and reel
motors could be required to regenerate to provide the torque needed for deceleration
to follow a speed reference profile. For the induction motor to regenerate, the speed
of the rotor must be greater than the speed of the flux-density wave that is produced
by its multi-phase power source. That is, the rotor speed must be greater than the
synchronous speed of the motor. In this case the rotor current, the polarities of the
rotor poles, and the torque direction are reversed from their values during motoring.
In Figure 6.6 for simplicity a torque-speed curve is shown for the generating mode at
nominal frequency and rated terminal voltage, assuming that the rotor speed is
changing such that its speed is greater than the fixed speed of the flux-density wave
which is set by the fixed stator frequency. While this differs somewhat from the mill
application wherein the frequency is continuously changing during a deceleration, it
nevertheless provides some feel for the motor characteristic during regeneration. In
this mode the magnitude of the breakdown torque (generating) is slightly higher than
the magnitude of the breakdown torque (motoring) as in Figure 6.5.

6.2.4 Variable Speed Operation

There are several methods to vary the speed of cage rotor induction motors. In this
section the techniques considered are those that are more applicable to control of

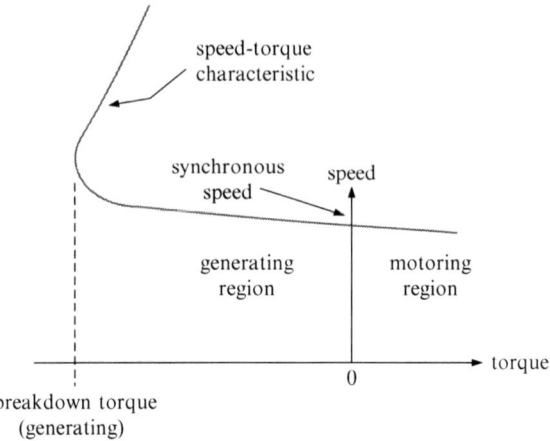

Fig. 6.6 Induction motor torque-speed characteristic, generating

the tandem cold rolling process and promote a basic understanding of the motor characteristics under the variation in the frequency of the multi-phase power source. The two areas that are considered are: (1) operation at and below base frequency, and (2) operation above base frequency. Operation at and below base frequency is essentially a constant torque area, while operation above base frequency is essentially a constant horsepower area. These two areas of operation are well-suited to the control of tandem cold rolling and are similar to earlier operation of the mill using direct current machines that exhibited a constant torque characteristic below base speed and a constant horsepower characteristic above base speed. Typical of these areas of operation are the mill stand drive motors listed in Table 6.1 where the horsepower and speed ranges for the drive motors of a typical tandem cold rolling mill are given. For example the drive motor of stand 1 is constant torque up to the base speed of 110 rev/min and constant horsepower above 110 rev/min up to 310 rev/min. Drive motors for the other mill stands are similar. Figure 6.7 depicts a torque-speed characteristic over the speed range of a typical mill drive motor.

The intent of what follows is to present the machine characteristics assuming a basic open-loop control technique which can provide a basis for the understanding of related methods and of some of the more complex control concepts.

Table 6.1 Typical mill stand drive motors	Stand	Motor horsepower	Motor speed (rev/min)
	1.	1,750	110/310
	2.	3,500	165/420
	3.	4,500	200/415
	4.	4,500	300/625
	5.	6,000	345/635

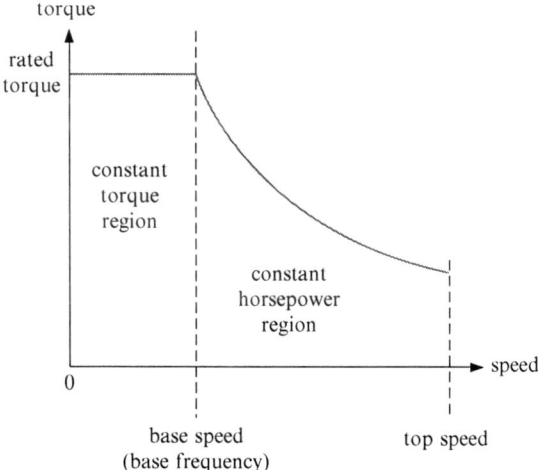

Fig. 6.7 Drive motor torque-speed characteristic

6.2.4.1 Operation at and Below Base Frequency

In a cage rotor machine the air gap flux is proportional to the air gap voltage divided by the frequency of the power source, *i.e.*,

$$\phi \propto \frac{E_{air\ gap}}{f}, \tag{6.13}$$

and

$$V_t = V_r + V_x + E_{air\ gap}, \tag{6.14}$$

where V_t is the machine terminal voltage, V_r and V_x are the voltage drops across the stator resistance and leakage reactance, and $E_{air\ gap}$ is the air gap voltage. At the base frequency, with rated voltage applied at its terminals the machine will run at its base speed, *i.e.*, the air gap flux is at its nominal value. To keep the torque-speed characteristic nearly invariant as the machine speed is reduced and to avoid higher magnetic saturation, the air gap flux must be kept constant. This implies that a reduction in the frequency must be accompanied by a proportional reduction in the air gap voltage. Over a major portion of the machine speed range the voltage drops V_r and V_x are small compared to V_t so that the terminal voltage is nearly equal to the air gap voltage. Thus if the machine speed is reduced by reducing the frequency, the terminal voltage also must be reduced proportionally. However, as the speed approaches zero the air gap voltage is reduced so that V_r and V_x are no longer negligible with respect to $E_{air\ gap}$ and the flux also is reduced. To compensate for this, the terminal voltage is boosted slightly at the lower frequencies to keep the flux approximately constant and thus retain the torque-speed characteristic at lower speeds. In this mode of operation the full torque of the machine is available as the speed is reduced, so that operation is essentially constant torque. Figure 6.8 depicts the torque-speed characteristics for motoring operation as the frequency is reduced below the base frequency. The characteristic for regenerative operation (Figure 6.6) also is repeated in a similar manner for frequency reduction below the base frequency.

6.2.4.2 Operation Above Base Frequency

Above base frequency the motor terminal voltage is held constant with the frequency being increased. The motor horsepower remains constant with the rated torque decreasing as shown in Figure 6.7. The air gap flux decreases according to (6.13). The torque-speed characteristic is as depicted in Figure 6.9. Operation in the regenerative mode above base frequency is similar, except that the individual torque-speed characteristic curves are for regeneration.

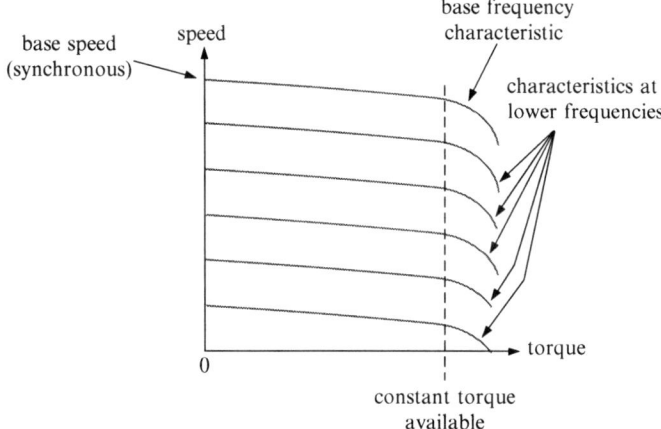

Fig. 6.8 Torque-speed characteristics of the cage rotor induction motor at and below base frequency

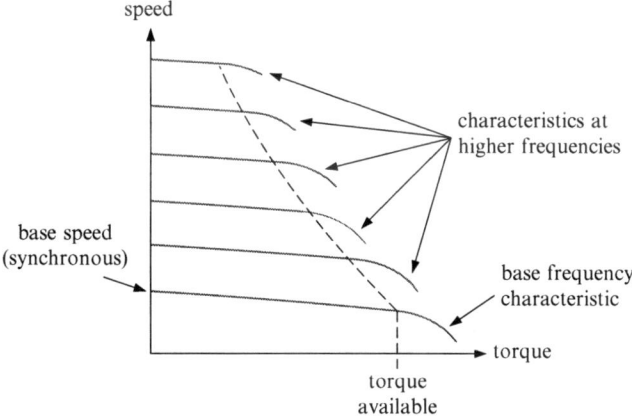

Fig. 6.9 Torque-speed characteristics of the cage rotor induction motor above base frequency

6.3   The Synchronous Motor

Synchronous machines are widely used throughout the world as generators of electrical power. In general, in industrial applications synchronous motors are used where a constant speed is desired. A desirable feature of the synchronous motor utilized in these applications is its capability to provide an adjustable power factor by adjusting the field excitation. Recently in many instances synchronous machines are being preferred over cage rotor induction machines as main drive motors for tandem cold rolling applications for reasons which are noted in Section 6.5.

The construction of the stator of the synchronous motor is similar to the stator of the cage rotor induction machine (Figure 6.1). The stator winding is placed in slots cut into the surface of the stator and is distributed to produce a nearly sinusoidal spatial mmf distribution around the air gap. The stator winding is denoted as the armature winding, as it is the winding in which a voltage is induced by the field of the rotating rotor. The stator winding is connected to the machine terminals for external connection to a multi-phase excitation source. The rotor of the synchronous motor is constructed with salient poles, or with non-salient poles which generally is similar to the construction of the cage rotor induction machine where the rotor has slots at its surface into which conductors are inserted. Unlike the induction motor, the field of the rotor of the synchronous motor is established by direct current excitation from a separate source rather than by induction via the rotating air gap flux. The excitation current is often supplied via a brushless excitation system wherein the exciter and associated rectifiers are mounted on the rotor shaft, which avoids the need for mechanical slip rings and brushes.

The non-salient pole rotor is also referred to as a cylindrical rotor. The cylindrical rotor results in a uniform air gap and has a distributed pattern of conductors placed in slots at the rotor surface to produce a sinusoidal mmf distribution with the same number of poles as the stator winding. For the main drive motors of tandem cold rolling mills non-salient pole rotors usually are preferred over rotors with salient poles due to their higher mechanical strength, a higher reliability, less maintenance, reduced windage losses due to their cylindrical-type construction, and a lower moment of inertia.

Similar to the cage rotor induction motor, the synchronous motor as applied to the control of the tandem cold rolling mill is capable of operation as a generator. During certain conditions of operation the synchronous motor can provide regenerative power to certain other motors in the mill, to the coilers, or to the plant power grid via the main drive associated with the motor. More detail regarding synchronous motors can be found in reference texts such as [1, 2].

6.3.1 Machine Torque

As in the case of the cage rotor induction machine, an understanding of how the flux and mmf waves interact to produce torque is essential to the understanding of the synchronous motor. To help develop a feel for the operation of the motor, it will be assumed that the motor is connected to a balanced three phase source of constant frequency so that there is a magnetic field in the air gap that rotates at synchronous speed as described previously in Section 6.2.1. It also is assumed that the motor losses and saturation effects are negligible, that the armature (*i.e.*, the stator) resistance and reactance are zero, that the rotor is turning at synchronous speed and that the rotor position and the field current (*i.e.*, the current in the rotor conductors) are established such that the voltages E_r induced in the armature by the current in the rotor are equal to the voltages of the three phase source, and that there are zero currents in the stator windings. Under these conditions the electrical angle δ_{RF} between the mmf wave F_r produced by the rotor current and the flux wave Φ_r in the air gap is zero, so that the rotor torque T_M is zero in accordance with the relation (6.11) as

$$T_M = K_T \left(\frac{P}{2}\right)^2 \Phi_r F_r sin(\delta_{RF}), \tag{6.15}$$

where P is the number poles and K_T is a constant.

It next is assumed that there is some load torque T_L subsequently applied to the shaft of the rotor, so that the rotor tends to slow down, the angle δ_{RF} is no longer zero as the position of the rotor is moving backward with respect to the position of the air gap flux wave, and rotor torque T_M is produced according to (6.15) in a direction so as to oppose the applied load torque (by Lenz's law). When the torques are balanced (*i.e.*, $T_M = -T_L$, after some transients) the speed of the rotor is again at synchronous speed, but its angle is no longer zero, and there are currents I_a in the armature windings which produce an mmf field A_a (denoted as the armature reaction) which adds to the mmf field of the rotor to produce a resulting mmf field R_t which produces the air gap flux Φ_r to keep the voltages induced in the armature equal to the source voltages. Figure 6.10 depicts the situations before and after the application of the load torque. For the simplicity of presentation, in this figure the rotating waves and fields are represented as phasors. If the load torque keeps increasing, the angle δ_{RF} will keep increasing until the angle is 90° which represents the maximum torque available at synchronous speed. The torque at the angle of 90° is denoted as the pull-out torque. The application of additional torque beyond the pull-out torque will cause the motor to slow down with the loss of synchronism. The torque-angle characteristic for a motoring application is as shown in Figure 6.11.

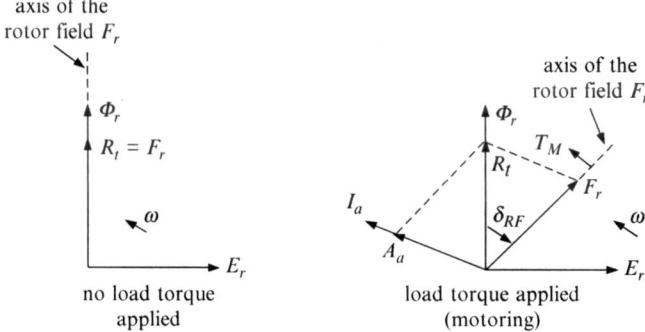

Fig. 6.10 Phasors for application of torque at the rotor shaft for a synchronous motor, motoring

6.3.2   Regenerative Operation

The conditions in the operation of the tandem cold rolling mill under which the synchronous motor can regenerate are as noted in Section 6.2.3 for the cage rotor induction motor. In these instances the load torque at the shaft is reversed and the machine acts as a generator. The phasor diagrams in this case are as shown in Figure 6.12. As in the motoring case the torque is limited by the pull-out torque in the generating region as depicted in Figure 6.11.

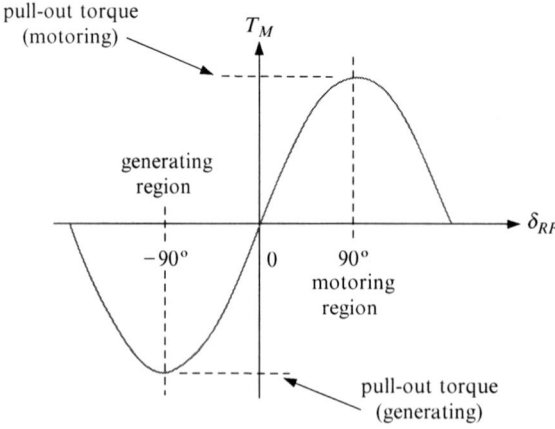

Fig. 6.11   Synchronous motor torque-angle characteristic

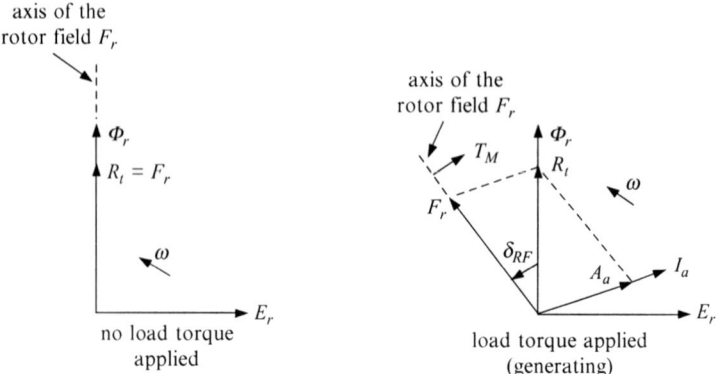

Fig. 6.12   Phasors for application of torque at the rotor shaft for a synchronous motor, generating

6.3.3 Variable Speed Operation

Variable speed operation of the synchronous motor will be addressed in this section by assuming that the voltage and frequency of the power source are varied, which is a typical method for control of tandem cold rolling mills. As in the case of the cage rotor induction motor, the areas of operation that will be considered are (1) operation at and below base frequency, and (2) operation above base frequency. What is presented assumes an open-loop control technique which provides an insight into machine characteristics as the voltage and frequency of the power source change, which helps with the understanding of more complex control methods.

6.3.3.1 Operation at and Below Base Frequency

It can be shown [3] that the machine torque also can be described as

$$T_M = K_T \frac{V_t}{f} \, sin(\delta_{RF}), \tag{6.16}$$

where V_t is the machine terminal voltage and K_T is a constant, and where in the derivation of (6.16) the resistive component of stator impedance is neglected at base frequency. To maintain constant torque as the frequency changes it is necessary to change the terminal voltage of the machine proportional to the frequency and also to increase the terminal voltage slightly to compensate for stator resistance as the frequency approaches zero, much the same as the compensation for stator impedance in the case of the cage rotor induction motor. The torque-speed characteristics for the synchronous motor when operated at and below base frequency

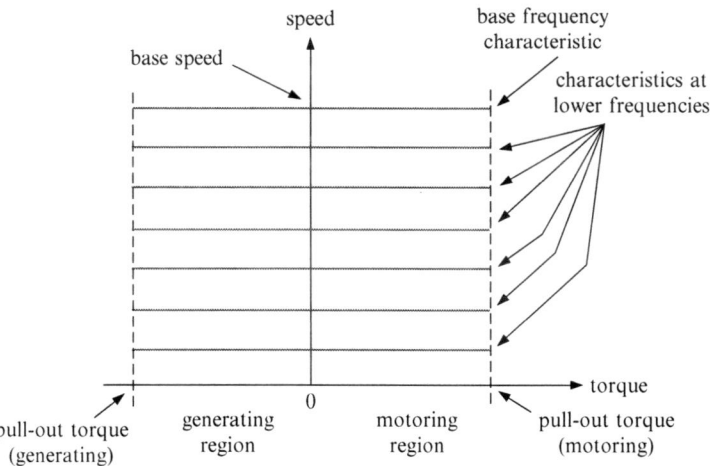

Fig. 6.13 Torque-speed characteristics of a synchronous motor at and below base frequency

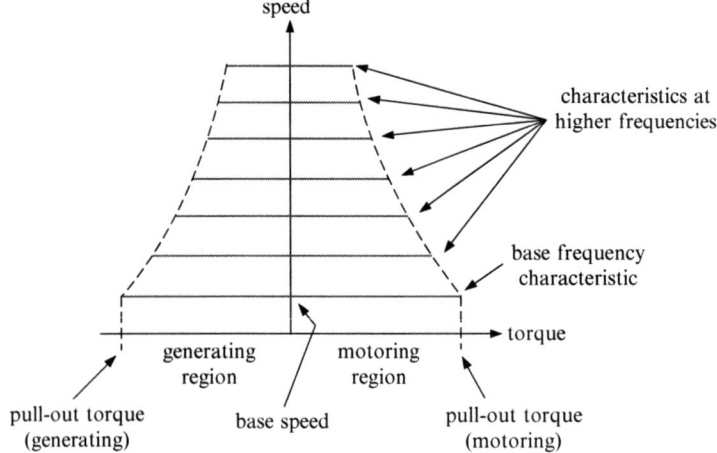

Fig. 6.14 Torque-speed characteristics of a synchronous motor above base frequency

are depicted in Figure 6.13 for operation in the motoring and generating regions of operation.

6.3.3.2 Operation Above Base Frequency

Above base frequency the terminal voltage is held constant as the frequency is increased. The horsepower in this region of operation is constant with the torque decreasing as the speed increases. The torque-speed characteristics are depicted in Figure 6.14.

6.4 Main Drives

This section describes the basic concepts of selected power electronic circuitry and certain control techniques associated with the main drives for tandem cold rolling mills. As with previous discussions, the intent of the material presented is to provide a background in fundamental concepts that can serve as a basis to aid the reader to understand the more advanced methods associated with the available modern technologies in this area. Certainly it is beyond the scope of this book to describe in specific detail the several methods available for the control of power from the plant power system to the variable speed drive motors, or the various techniques that can be used for control of the drive itself. What is presented will concentrate on the main concepts that underlie most of the viable modern methods that are presently in use and have proven to be successful in several recent applications, with more detail available in the quoted references and other associated literature. What follows will assume that the motor to be controlled is either a synchronous machine with a cylindrical rotor or a cage rotor induction motor, and that the controller is a voltage source converter with an active front end. The use of a cycloconverter as an alternate method also will be briefly addressed. The control will be assumed to be a closed-loop field oriented (*i.e.*, vector) control that is suitable for these types of machines.

6.4.1   Motor Drive Power Circuitry

The preferred circuitry that controls power to the motor is depicted in Figure 6.15, which is typical of modern equipment that is operating successfully in several applications. As can be seen in Figure 6.15, there are three main parts which comprise the power circuit of the drive: (1) the active front end, (2) the DC link, and (3) the motor inverter.

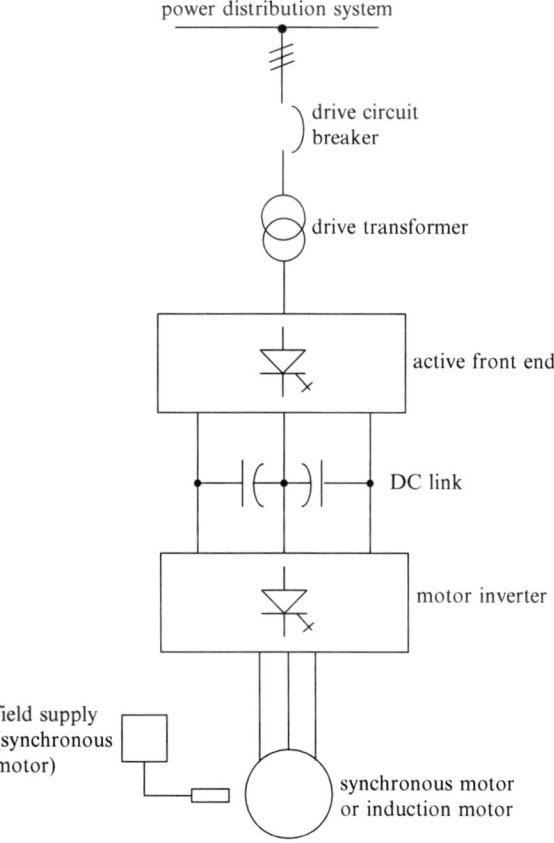

Fig. 6.15 Voltage converter drive, power circuit schematic

6.4.1.1 Active Front End and DC Link

The active front end is a bridge arrangement of power semiconductors that provides power to the large capacitors in the DC link. In most modern applications the semiconductors are IGCTs (integrated gate commutated thyristors). IGCTs are fully controlled thyristors which can be switched on and off by a gating signal. The active front end is a converter which provides for full motoring and regenerative capability of the drive. The gating of the IGCTs is controlled such that the line current is very nearly sinusoidal so that harmonics injected into the power system are significantly reduced. In addition, the DC link voltage is kept essentially constant which decouples the motor inverter from the plant power system so that the drive system is less sensitive to line voltage fluctuations and can even tolerate brief dips in the line voltage. The reactive power also is controlled so that a power factor of unity can be realized, or power factor correction can be provided to compensate for other loads connected to the same power source as the drive.

The DC link consists of large capacitors to provide a very stiff voltage source to support the proper operation of the motor inverter. The capacitor circuit is split as shown in Figure 6.15 to create a neutral point which is necessary for proper inverter operation.

6.4.1.2 Motor Inverter

Similar to the active front end the motor inverter consists of a bridge arrangement of power semiconductors to control the power provided to the motor. The motor inverter is more generally denoted as a voltage source converter as the power circuit can operate as a controlled rectifier in one direction, and in the other direction as an inverter which converts the voltage of the DC link to a waveform suitable for variable speed operation of the drive motor, to provide full motoring and regenerative operation. In this case however operation is mostly as an inverter and hence the notation of inverter is more appropriate. The inverter uses IGCTs in a three level three phase configuration with a neutral point provided in the DC link by the split capacitor arrangement. More detail related to the operational detail of the inverter can be found in [4].

The inverter is pulse width modulated (PWM). In the pulse width modulation method a square wave is broken up into pulses of varying width to adjust the magnitude of the output fundamental. Figure 6.16 provides an example of a pulse width modulated waveform for the output voltage V_{ab} of a PWM inverter with a fixed voltage source input of V_d and a sinusoidal fundamental output voltage.

An advantage of PWM over other methods (such as the six step inverter) is that bulky and expensive filters to reduce lower order harmonics that cause large distortions in the current waveform are not needed. With the PWM technique the IGCTs in the inverter are controlled to significantly reduce the harmonics and at the same time control the output voltage. There are several methods of PWM the specifics of which are discussed further in [4].

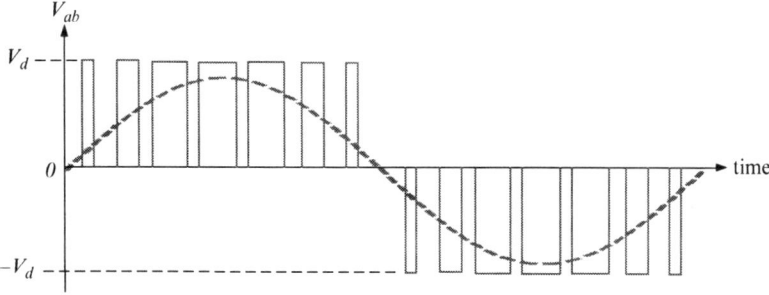

Fig. 6.16 PWM voltage output waveform, with sinusoidal fundamental

The efficiency of the inverter is very high as the commutation losses in the inverter itself are quite low which supports the overall efficiency of the drive of about 99%. Because the output current of the inverter has very low total harmonic distortion, the torque pulsations at the motor shaft are reduced to very low values, and there is negligible derating of the drive motor, all of which make the PWM inverter attractive as a method of controlling motor speed and torque.

6.4.1.3 Cycloconverter

The cycloconverter is a less preferred drive for the main mill motors and is included for completeness as it remains available as an alternate method. Figure 6.17 is a schematic of the cycloconverter circuitry for a typical system.

The cycloconverter is a frequency changer that converts power at one frequency to power at another frequency in a single stage process. Cycloconverters are generally used for very large drives in various industrial applications. In the case of tandem cold rolling the cycloconverter drive provides adjustable frequency and voltage to a main drive mill motor by varying the outputs of dual converters which control the individual phases of the drive motor. A dual converter is a power bridge of thyristors which provides a DC output with an alternating current input from the plant power system. In the cycloconverter the output of each dual converter is controlled to vary the DC output such that a nearly sinusoidal current at a frequency lower than the input frequency is provided. Cycloconverters are limited in the maximum output frequency that can be attained and still maintain a current waveform that is reasonably sinusoidal. Typically the maximum frequency of the output is about 50% of the frequency of the input.

Cycloconverters can be controlled so that the output current in the drive motor has a low harmonic content in the motor torque. However, at the input the harmonic content consists of non-integer harmonics which can be difficult to filter, even though the harmonic content is somewhat reduced below that of a DC drive because the firing angles of the thyristors in the dual converters are continuously changing. Moreover, since the maximum frequency of the cycloconverter is much less than the maximum

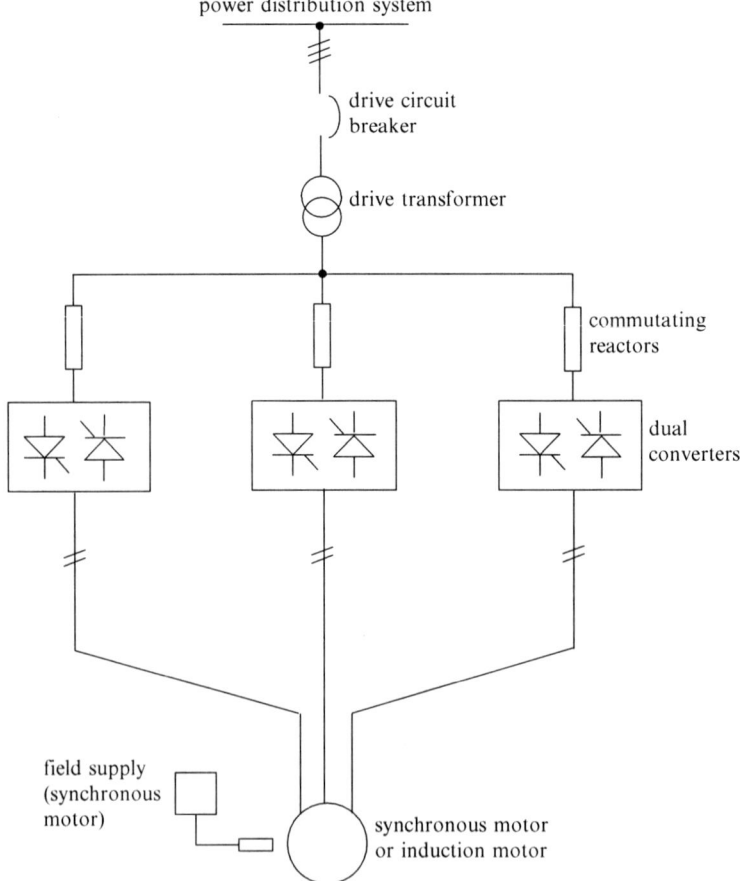

power distribution system

drive circuit
breaker

drive transformer

commutating
reactors

dual
converters

field supply
(synchronous
motor)

synchronous motor
or induction motor

Fig. 6.17 Cycloconverter drive, power circuit schematic

frequency of the voltage converter, a physically larger drive motor must be used which increases the cost. The maximum power factor of a cycloconverter drive is at best 0.8, and is reduced as the frequency and voltage are lowered. While the efficiency of the cycloconverter can be as much as 99% its lower power factor, higher harmonic content at the line side, and the requirement for a larger motor make it somewhat less attractive than the voltage converter as a main drive. Additional material describing the operation of dual converters and cycloconverters can be found in [4, 5].

6.4.2 Drive Controller

Modern drive controllers are based on the principle of field oriented (or vector) control. A very significant advantage of vector control is that when applied to a cage rotor induction motor or a synchronous motor, the ideal performance characteristics are nearly identical to those of a separately excited DC motor which has been used quite successfully in earlier applications for the control of tandem cold rolling mills. The field oriented control also is denoted as orthogonal or decoupling control, the reasons for which will become obvious in the material that follows. To better understand the benefits of vector control it is helpful to briefly review the characteristics of the separately excited DC machine.

In the DC motor the torque T_M can be expressed as the product of the shunt field current and the armature current, assuming that saturation and armature reaction can be neglected, *i.e.*,

$$T_M = K \, \phi_f I_a, \tag{6.17}$$

where ϕ_f is the flux created by the machine shunt field current, I_a is the armature current and K is a constant. In the machine the flux created by I_a is orthogonal to ϕ_f and both fields are stationary in space. Because of the orthogonality, there is decoupling between the shunt field flux and the armature current, so that a fast response in the machine torque can be achieved by adjusting the armature current without affecting the field current, and if the field current is adjusted the armature current remains unaffected. Using appropriate control methods it is possible to extend these desirable characteristics to the cage rotor induction motor and the synchronous motor.

6.4.2.1 Field Oriented Control of the Cage Rotor Induction Motor

In the techniques for scalar control of the cage rotor induction machine there are interactions between the torque and the flux as both depend on frequency and the voltage or current. For example, if the torque is increased by changing the frequency, the flux will tend to decrease in a somewhat sluggish manner. The decrease in flux can be compensated by a control loop which then raises the voltage, but the response remains sluggish and the torque sensitivity with respect to slip is reduced.

In the 1970s the concept of field oriented control (or vector control) was developed to overcome this shortcoming. The idea of vector control is based on the use of a transformation of machine variables into a reference frame which rotates at synchronous speed, and in which the flux and current vectors are expressed in an orthogonal framework. Thus it is possible through appropriate coordinate changes to control the stator currents in such a manner that the variables as expressed in the rotating reference framework react to produce torque similar to the DC machine, *i.e.*, the torque produced as expressed in the rotating reference frame is

$$T_R = K \, \phi_r I_q, \tag{6.18}$$

where ϕ_r is the peak value of the rotating machine flux vector that is produced by the flux producing component I_d of stator current I_s, I_d is in phase with ϕ_r, I_q is the component of the stator current that is orthogonal to ϕ_r and I_q, the increase of which increases the torque without changing ϕ_r, and K is a constant. Figure 6.18 depicts the variables as they appear in the synchronously rotating framework, and shows how an increase in the torque producing component of the stator current increases the torque without affecting the flux, and how an increase in the flux producing component of the stator current increases the flux without affecting the torque component of the stator current. In Figure 6.18 it is assumed that the rotor reactance can be neglected so that the voltage E_r resulting from the change in the flux is also the voltage across the rotor resistance.

A schematic of a functional implementation of a method of field oriented control is shown in Figure 6.19. The specific details of the reference frame transformation, the generation of the appropriate signals for its control, the estimation of the flux vector ϕ, and other methods of vector control are similar to those described in [4]. Because the control of torque does not affect the flux, the transient response of the vector controlled drive is very much like the response that would be obtained with a DC machine and is quite suitable for drives for tandem cold metal rolling using cage rotor induction machines.

6.4.2.2 Field Oriented Control of the Synchronous Motor

As in the case of the cage rotor induction motor, the transient response of a synchronous machine that is driven from a scalar controlled drive can be significantly improved by the use of field oriented control. In the case of the synchronous machine, this is mainly due to the capability of the field oriented controller to compensate for the slow rise in the rotor field current by adding magnetizing current to increase the flux during transient conditions.

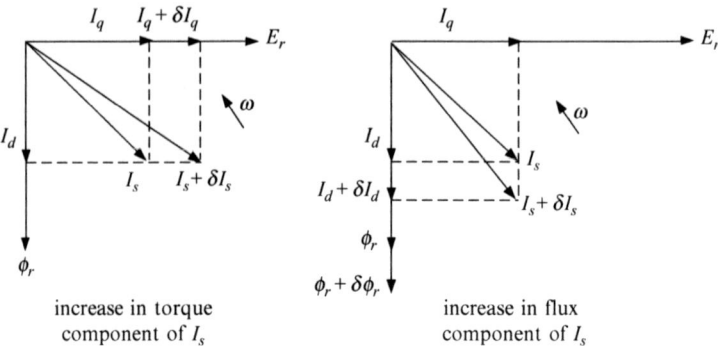

Fig. 6.18 Phasor diagrams, showing effects of changes in torque and flux components of stator current in an orthogonal rotating framework

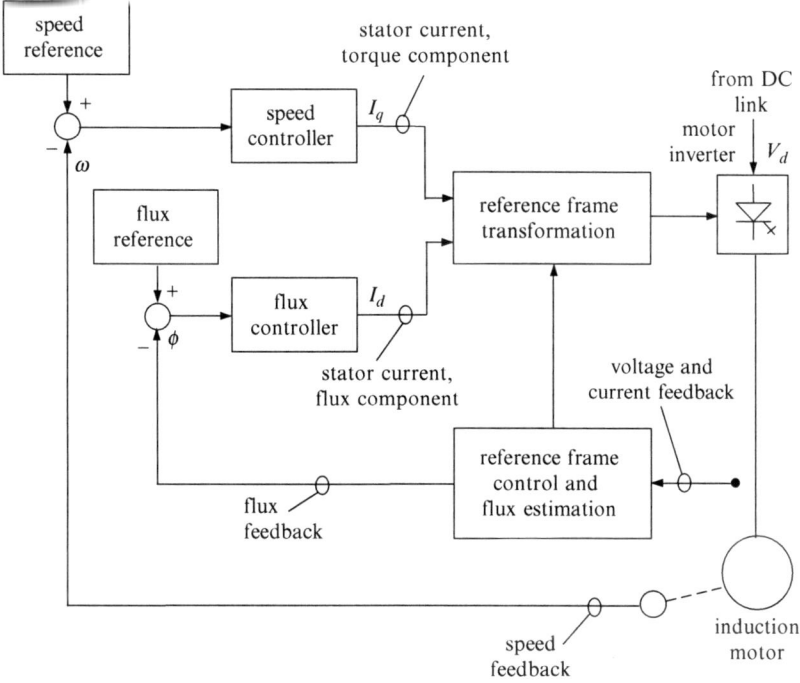

Fig. 6.19 Schematic of a field oriented controller for a cage rotor induction motor

The field oriented drive for the synchronous motor is more complex than for the cage rotor induction machine, although there are some similarities. Figures 6.20 and 6.21 are schematics of the implementation of such a drive.

The concept of transforming machine variables into a reference frame that rotates at synchronous speed, and in which the flux and current variables are expressed in an orthogonal framework is similar to what is done for control of the cage rotor induction motor. The main difference is in the control of the rotor field current. In the case of the synchronous motor at steady-state the flux is set by the current in the rotor windings. During transient conditions which could be caused by perturbations to the system or changes in the references, the slower response of the rotor field current causes the response of the controller to be slower. Referring to Figure 6.21, at steady-state the magnetizing current I_m is related to the rotor field current I_f by the angle δ as

$$I_m = I_f cos(\delta), \tag{6.19}$$

and the reference to the field current controller I_f^* is related to I_m^* as

$$I_f^* = \frac{I_m^*}{cos(\delta)}. \tag{6.20}$$

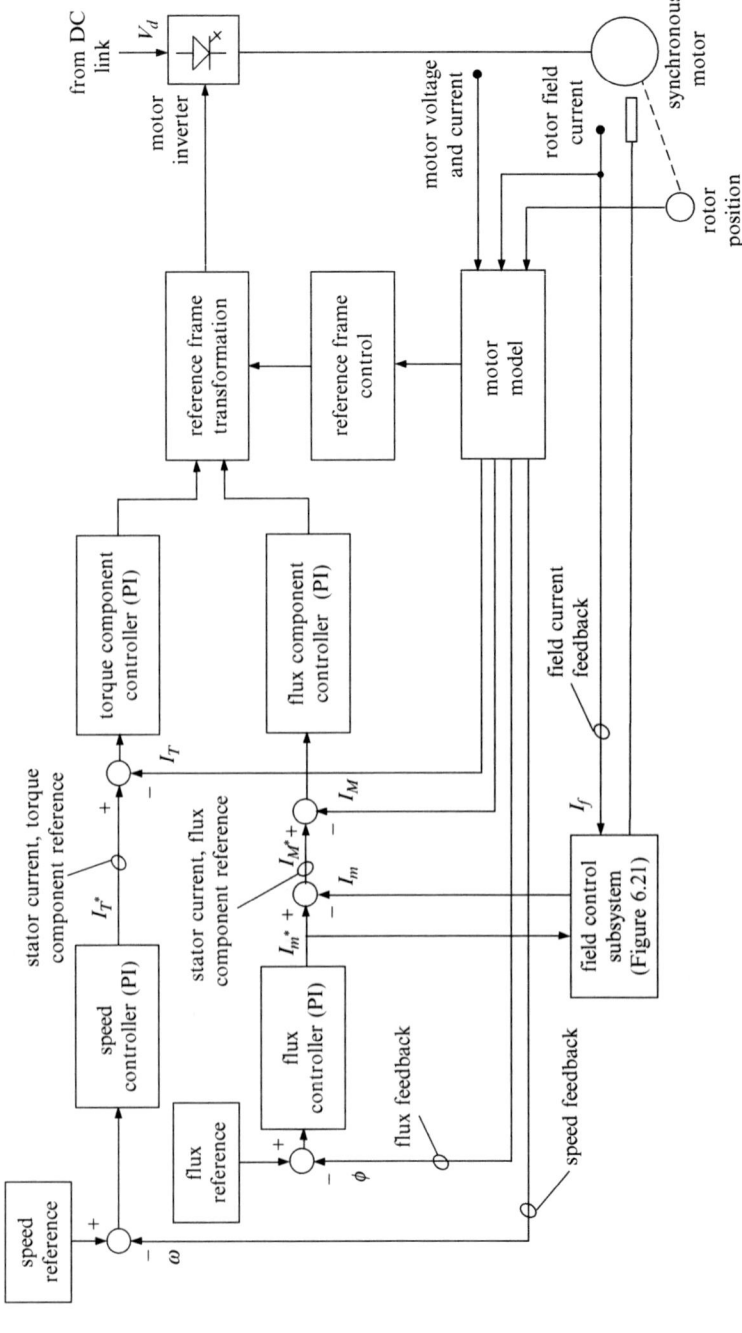

Fig. 6.20 Schematic of a field oriented controller for a synchronous motor

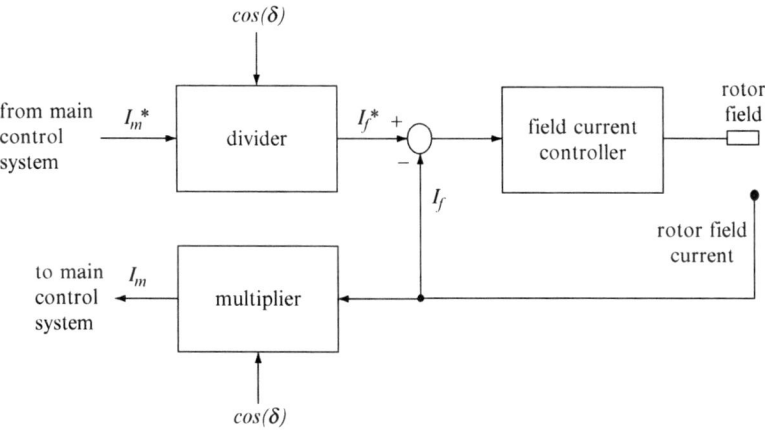

Fig. 6.21 Schematic of field control subsystem (supplements Fig. 6.20)

At steady-state the synchronous machine is operated at unity power factor so that the angle between the stator voltage and the stator current is zero. The torque component of stator current is represented as I_T and the flux component I_M of stator current which is orthogonal to I_T is zero. Under this condition, the flux is established by the rotor field current I_f so that in Figure 6.20 the reference I_m^* is balanced by I_m which is supplied solely by the rotor field current I_f as in (6.19). This makes the reference I_M^* zero and by closed-loop control action makes I_M zero.

Under transient conditions, as for example with the machine operating in the constant torque region (constant flux reference) a sudden load torque disturbance occurs which causes an increase in the angle δ. As shown in Figure 6.21, the increase in δ causes an increase in the field current reference I_f^* with I_m^* remaining initially fixed. The field current I_f begins to increase slowly, with I_m initially decreasing and then beginning to increase slowly. I_M^* is no longer zero and adds a flux component to the stator current which in turn changes the flux to attempt to compensate for the changing field current. After the transient is over, the flux is at its initial value, but during the transient the slow change in I_f is compensated for by a faster change in the flux component I_m of the stator current, with the machine returning to a unity power factor condition and I_M^* again at zero. Operation in the field weakening region is similar. Thus the beneficial effect of the field oriented control in improving the response can be recognized and confirms the usefulness of this technique in control of synchronous motors for tandem cold rolling.

6.5 Motor and Drive Selection

The selection of a suitable motor and drive for the tandem cold rolling process is a task that is essentially application specific. Both the cage rotor machine and the synchronous motor have some advantages and disadvantages which should be carefully considered. For example, in general the synchronous motor is more efficient than the cage rotor machine, and if supplied with a cylindrical rotor has a lower moment of inertia which may be advantageous for control. On the other hand the synchronous machine generally has a higher initial cost compared to the induction motor, and also could result in a higher installation expense.

The cycloconverter drive is less preferred over the voltage converter drive for the reasons previous noted. In both the cycloconverter and the voltage converter the preferred method of control is a field oriented (vector) control technique.

Many of the details of issues that are related specifically to motors and drives often can best be determined based on consultation with the motor and drive manufacturers. In addition, other associated issues such as maintenance costs and the availability of supporting services from equipment manufacturers should not be overlooked. Except in certain special cases, the motor and its associated drive should be evaluated as a single entity and obtained from the same manufacturer to assure compatibility of systems and hardware, and to avoid the likelihood of having to resolve issues related to interfacing that could arise if the motor and drive were obtained from different manufacturers.

References

1. Fitzgerald AE, Kingsley Jr C. Electric machinery the dynamics and statics of electromechanical energy conversion. New York: McGraw-Hill; 1961.
2. McPherson G, Laramore RD. An introduction to electrical machines and transformers. New York: Wiley; 1990.
3. Sen PC. Principles of electric machines and power electronics. New York: Wiley; 1967.
4. Bose BK. Modern power electronics and AC drives. Upper Saddle River: Prentice-Hall PTR; 2002.
5. Pelly BR. Thyristor phase-controlled converters and cycloconverters; operation, control, and performance. New York: Wiley Interscience; 1971.

Index

J. Pittner and M.A. Simaan, *Tandem Cold Metal Rolling Mill Control*,
Advances in Industrial Control, DOI 10.1007/978-0-85729-067-0,
© Springer-Verlag London Limited 2011

Other titles published in this series (continued):

CPSIA information can be obtained at www.ICGtesting.com
225754LV00006B/39/P

9 780857 290663

	DATE DUE		